PHYTOGEOMORPHOLOGY

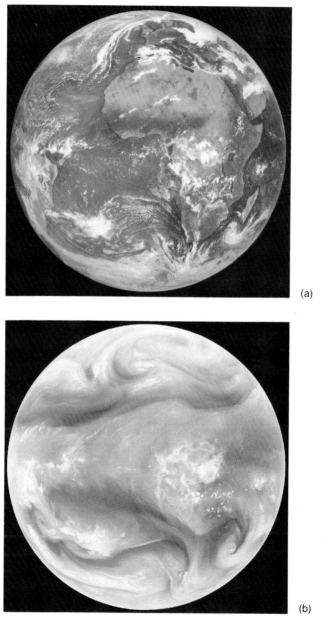

(a)

(b)

Figure 2.1. Simultaneously recorded Meteosat imagery, taken at 1155 hr GMT on 26 March, 1982, recorded and processed at the European Space Observation Centre (ESOC), Darmstadt: (*a*) scene recorded in the visible spectrum showing Africa partly cloud covered but with some vegetation pan formations visible; (*b*) scene recorded showing troposphere water vapor in a series of whorls, anticlockwise in the northern hemisphere and clockwise in the southern hemisphere.

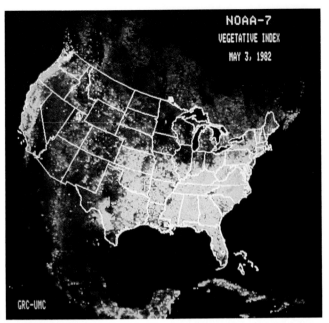

(a) May 3

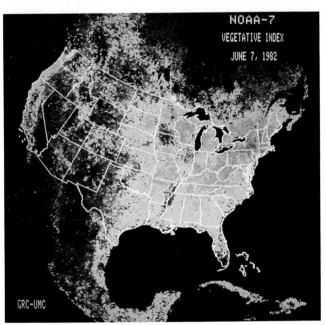

(b) June 7

Figure 8.2. Seven-day color composites of vegetation greenness index derived from NOAA AVHRR data recorded in the visible and near-

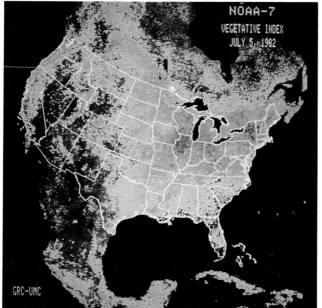

(c) July 5

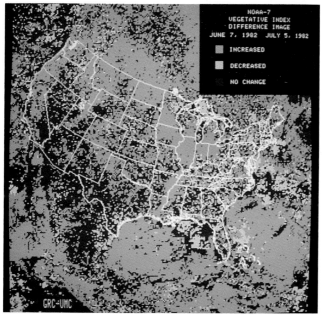

(d) Greenness change between June 7 and July 5

IR bands. Note the seasonal differences in vegetation greenness of the United States between May and July, 1983.

PHYTOGEOMORPHOLOGY

J. A. HOWARD
C. W. MITCHELL

A Wiley-Interscience Publication
JOHN WILEY & SONS
New York • Chichester • Brisbane • Toronto • Singapore

Library of Congress Cataloging in Publication Data:

Howard, J. A.
  Phytogeomorphology.

  Bibliography: p.
  Includes index.
    1. Botany—Ecology.   2. Phytogeography.   3. Geomor-
phology.   4. Plant communities.   I. Mitchell, C. W.
II. Title.
QK901.H69   1985   581.9   85-3218
ISBN 0-471-09914-7

Printed in the United States of America

10 9 8 7 6 5 4 3 2 1

To

M.B.H. and A.C.S.M.

PREFACE

This book brings together those aspects of the study of vegetation and geomorphology that can be readily synthesized in the study of the landscape, defined broadly as the portion of land or territory on the earth's surface that the eye can comprehend. The intention is to show how these two factors of the landscape in combination not only to some extent govern other environmental factors, but also lead to the articulation of small-scale, relatively homogeneous subdivisions of the earth's surface for the purposes of planning and management. In doing so we have often drawn on personal observations in the many countries visited, and an attempt is made to provide a wide-scale view of the subject. This, we hope, will encourage others to develop regional in-depth studies.

The book is addressed to all those concerned with the geographical relationships of the earth's surface, particularly to the professional agriculturist, engineer, forester and environmentalist, to the student studying natural resources, and to the planner whom an integrated view of these relationships can help in decisions about land management. More generally, the book will also help many nonspecialists who are interested in an understanding of landform–plant relationships.

There are four ways in which we seek to meet these ends. First, we present the theory of phytogeomorphology in the hope that in the future it will be taught increasingly as a single subject. Second, we hope to encourage others to increase the use of plant–landform relationships in their work, so that the geomorphologist will place more reliance on vegetation and the plant ecologist more reliance on geomorphology. Third, we emphasize plant–landform relationships in practical studies of the natural environment, not only because they express many of the key variables determining dynamic processes, but also because these two factors are the ones that are most readily seen and measured. Fourth, we believe that combining these two aspects of the environment has considerable educational and practical merit. In particular, the combination underlines the interrelations among the natural features of the landscape, strengthens the understanding of these relationships, and can lead to an integrated view, which is fundamental to planning land uses and evaluating potential land uses.

The book begins with introductory chapters on the conceptual background and on climate as it provides the first broad subdivision of the earth's landscapes. It then approaches the subject by reviews of geomorphology (Chapter 3) and Vegetation (Chapter 4), which are then integrated by studies of the landscape at different scales (Chapters 5 and 6). Thereafter, landforms and vegetation are considered simultaneously in relation to the use of modern remote sensing techniques (Chapters 7 and 8), field survey (Chapter 9), thematic mapping of the results of surveys (Chapter 10), and the application of methods of data processing (Chapter 11). The practical applications of phytogeomorphology are examined broadly in terms of actual and potential land uses (Chapter 12), and then with reference to specific discipline-oriented studies (Chapter 13). Finally, both theoretical and practical conclusions are drawn (Chapter 14).

The text expresses only our personal views and we wish to make the following acknowledgments of permission to reproduce illustrations: the European Space Agency, Figure 2.1; the Royal Meteorological Society, London, Figure 2.2; the Food and Agriculture Organization of the United Nations, Figures 2.3, 8.3, and 13.2; L. J. Holdridge, Figure 4.2; *Geoforum* magazine, Figures 5.1, 6.1, 6.2, and 6.4; Zeiss, Carl Zeiss, Oberkochen, Figure 7.2; U.S. National Aeronautics and Space Administration EROS Data Center, Figure 8.4; Reading University, Figure 8.5; and the Sudan government, Figure 13.1. Figure 8.9 is used by courtesy of the U.K. Ordnance Survey and is Crown Copyright. Figure 10.3 is from *Principles of Cartography* by E. Raisz, copyright 1962, McGraw-Hill Book Company, and is used with kind permission of the McGraw-Hill Book Company.

J.A. HOWARD
C. W. MITCHELL

June 1985

CONTENTS

13   Applications of Phytogeomorphology   179

14   Conclusions   198

PHYTOGEOMORPHOLOGY

ONE

INTRODUCTION

1.1 THE IMPORTANCE OF LANDFORM–PLANT RELATIONSHIPS

Excluding those creatures that live wholly on the produce of the seas, all life depends directly or indirectly on a very thin mantle of the earth's surface. This is closely associated with the landscape and the form of the land and the soil, which in turn reflects the operation of climate on the vegetation and physical materials of the earth's surface.

Landscape can be defined as the portion of land or territory that the eye can comprehend, including all the objects it contains. Landform and vegetation are generally the main features because they constitute the two most observable elements (Figure 1.1), because they reflect the influence of the other important environmental factors, and because they complement each other timewise in the development of landscape. They are the context of everyday life, visible from our windows at home and work, and surrounding us on our journeys. They influence the daily and long-term decisions of planners, farmers, foresters, engineers, environmentalists, and many others. They dominate the esthetic quality of our rural environment. Even our choice of home and holiday locations is often arranged to optimize the advantages of the landform and vegetation.

At the broadest scale of landforms, the locations of the major continents on the earth's surface govern the distribution of continental, maritime, and monsoonal climates, and the paths of ocean currents with their associated effects on the climates of coasts exposed to them. The earth's major temperate and tropical forests, initially located by these factors, contribute to the earth's climatic balance by their direct effect on the albedo of its surface and the large-scale ab-

1

Figure 1.1. Landforms and vegetation are the major observable features of the natural land-scape. The Agri valley, Basilicata, Italy, shows rivers incised into Pleistocene conglomerates. Hill-tops support cereal cultivation, steeper slopes are covered with oak woodland (*Quercus pubescens*), and river valleys with stony floors support scattered low shrubs (i.e., *Rosmarinus* sp.).

sorption of carbon dioxide, transpiration of moisture, and emission of atmospheric oxygen.

At mesoscale, these interactions are equally intimate, with each component influencing and being influenced by others. The distribution of highland and lowland within a continent owes its origin to tectonic forces, but these may in their turn have been partly due to isostatic adjustments to past climatically induced sedimentation. For example, the tectonic activity that has formed the Mediterranean mountains was derived from the overloading of the Tethys geosyncline with materials eroded from the adjacent Laurasian and Gondwana shields to north and south, respectively. Once mountains are formed they act as a climatic control, especially where they are large enough to deflect or block the path of global wind circulations. The Himalayas effectively shut off central Asia from India, turning the former into a desert and exposing the latter to the more intensive heating that speeds and intensifies the monsoonal rains. Similarly, the Andes cause the Patagonian desert by blocking off the rain-bearing westerly winds.

Conversely, climate in the long term molds landforms by the processes of weathering, erosion, and sedimentation, which, although ubiquitous, nevertheless have different characteristics in different rainfall and temperature zones. For example, the vast alluvial Sudan plains south of Khartoum are mainly formed of outwash from the Ethiopian highlands brought down by the annual monsoon. In colder climates or at high altitudes the spring thawing of snow cover causes accelerated erosion and sedimentation. This process has contributed largely to the deposition of such wide alluvial plains as those of the Indus, Ganges, Tigris, Euphrates, Mississippi, and Po, and the wide detrital aprons that parallel the mountainous spinal cord of the Americas.

Worldwide vegetation distributions clearly mirror these variations of climate-induced topography and topographically induced climate, but themselves impose modifications in return. Plant communities influence local climate not only in the ways previously mentioned, but also at the microlevel by intercepting rainfall and solar radiation and by slowing winds and increasing their turbulence. The presence of vegetation favors increased relative humidity and reduced daily and annual temperature variations.

Vegetation modifies landforms through its effect on soil. By penetrating into rock cracks, plant roots increase weathering and convert parent material into the mantle of unconsolidated surface material called the *regolith*. By binding the soil surface, they keep it in place. Thus their total effect is to increase soil depth and diminish its movement. By providing organic matter they assist soil structure and aeration and facilitate the infiltration and circulation of groundwater. This has the effect of diminishing surface runoff and erosion and of smoothing landscapes. Hills assume a more even convex–concave profile rather than the sharp angles and bare slope forms that characterize the scenery of deserts or polar regions. For example, this contrast is illustrated by a comparison between the smooth forest-clad slopes of the California Coast Range and the bare angular profiles of the nearby Mojave Desert.

Landform–plant relationships greatly influence and reflect several major environmental factors. The most important of these are water, soils, and animals. The water balance in any environment is a function of the way the rainfall is distributed. The ratio between runoff and infiltration depends directly on the slopes and permeability of the ground surface. Because plants, in addition to their substantial consumptive use, transpire much more water than bare surfaces, they are, except in the arid and polar regions, the dominant influence in determining the total evapotranspiration.

The importance of the plants and landforms in the genesis of soils is summed up and well illustrated by Jenny's factorial equation (1941)

$$S = f(c, p, r, o, v)t$$

In this equation, S = soil profile characteristics, c = climate, p = parent materials, r = relief, o = organisms including animals, v = vegetation, and t =

elapsed time. Changes in any variable will be reflected in the soil character. Because climate is mainly significant through its effect on plants and because both surface fauna and soil microorganisms are chiefly dependent on the decay of plant tissues, the main causative factors are relief, parent material, and vegetation.

Surface fauna likewise depend greatly on the density and quality of the vegetal cover. Although animals are clearly not fixed in location in the same way as plants, they are generally, especially at the lower trophic levels, so intimately involved in ecosystems that they form part of it. This involvement is also true of human beings not only because they use the landscape but also because they modify it in a multitude of ways. They plough, dig, and level its surface, bore and tunnel underneath it, dam its rivers and develop it with buildings and roads. They alter the plant life to such an extent that in much of the temperate world the vegetal cover is mainly induced and only skeletal natural ecosystems exist. Long (1974) quantified this process of modification by categorizing the human interference in the natural landscape on a numerical scale of "artificialization" from deserts at one extreme to highly urbanized environments at the other.

A final reason for emphasizing the combination of landforms and vegetation is that both are directly and readily recognized and recorded by modern methods of resources survey, which include the increasing use of remote sensing. The past 20 years have seen a great increase in the variety and speed of resources surveys. Most of the satellite imagery of the land surface of the earth is primarily a record of terrain and vegetation types so that these two factors have become the basis for the interpretation of the many less clearly visible factors of the earth including soils and geological structures.

1.2 DEFINITIONS

The term *phytogeomorphology* will be used to emphasize the importance of combining two major environmental factors—plants and landforms—in studies related to the land surface of the earth and in recognition of their interdependence. Whereas the present-day landforms usually predate history, existing plant communities reflect the interaction of recent and current environmental factors. Thus, in combination, the two form a powerful tool for the survey, management, and planning of our environment. If the reader is a plant ecologist or geomorphologist, there is probably little new in the following text on one's specific discipline, but what is important is to appreciate its relationship to the other discipline and to use the two disciplines in combination.

Phytogeomorphology imbues, through the study of the vegetation, aspects of the population–environment interaction that are expressed in the form and nature of the ecosystem. Its central concept reflects those sensitive landform–vegetation relationships that are visibly dominant on the landscape. Phytogeomorphology excludes consideration of the stratosphere and upper troposphere, the oceans, except for those parts of their inshore coastal beds that sustain appreciable plant life, and the deeper subsoil of the lithosphere beyond the range of plant roots. The works and activities of animals and humans are significant in so far as they directly cause or result from the plant–landform relationship.

The history of related terminology is also of interest in helping to place phytogeomorphology in perspective and shows the frequency with which it has been necessary to coin new words and combinations of words to express the earth–surface relationships and concepts. This list of terms is long and the distinctions between them are often subtle and may not always be clear cut. The more important terms will now be mentioned.

One of the earliest significant words is *physiography*, first introduced by Linnaeus, which, although less used today and succeeded partly by the term *biogeocoenose*, has retained the wider meaning of the study of all the earth's exterior features, including atmosphere and oceans as well as living things. *Geomorphology* is narrower and means the study of the origin and systematic development of all types of landforms, including their relief. *Relief* has a still narrower meaning and can be defined as the elevations and surface undulations of the landscape.

Geobotany originally referred only to traditional plant taxonomy, but around 1800 Alexander von Humboldt expanded its use to cover the discipline of plant geography, and in Europe the term has come to be applied to field botany (Grisebach, 1872; Rübel, 1930; Mueller-Dumbois and Ellenberg, 1974), and the study of communities especially in relation to the higher levels of organization (Dansereau, 1951). In the past few decades the term has also been used for the science concerned with the recognition of geological phenomena, particularly mineral resources, from their plant cover (Sukachev and Dylis, 1966; Brooks, 1983). There is no exact Anglo-American equivalent to the original European use of the term, but there are practical synonyms for its subdivisions (Mueller-Dumbois and Ellenberg, 1974).

The term *plant geography* was used by Schimper (1903) and Warming (1909) and *phytogeography* by Croizat (1952) as general terms for the geography of plants. Although retained in this sense in Anglo-American literature, the common European term is now *floristic geobotany*. The term *ecology* was first suggested by Haeckel in 1866 to signify the study of plant–habitat relationships, which is derived from the Greek word *oikos*, a house. This was subdivided later by Schroter (quoted by Troll, 1971) into *autecology* and *synecology*, depending respectively on whether it was a study of habitat conditions in relation to single plant species or to entire plant communities. Synecology is today regarded as equivalent to a range of terms including *vegetation ecology, plant sociology*, *phytosociology*, and *phytocoenology*. American and British publications use the term *ecology* as generally equivalent to synecology, but they extend it to include animal as well as plant ecology. *Physiological ecology* has sometimes been used as an equivalent of autecology.

Troll also suggested the term *Landschaftsökologie* (landscape ecology) and as early as 1939 advised that the science it represented be used in conjunction with the interpretation of aerial photographs as a means of exploring little-known landscapes, and that this would be one of the most important directions for geographical research in the future (1971).

Plant ecologists have also varied greatly in defining community (cf. Greig-Smith, 1964). It may be sufficient to demonstrate these differences of opinion by saying that, on the one hand, a plant community has been viewed as a complete organism or semiorganism (Clements, 1916; Tansley, 1935); and, on the other hand, simply as an assemblage of plants. Beadle and Costin (1952) viewed a community simply as "an assemblage of plants" but included "and their related fauna." The study of the distribution of plant species dates back to the time of the early systematists and Linnaeus.

From the broader phytogeographical standpoint, plant ecology was also termed *ecological plant geography*. Warming (1909) subdivided *plant geography* into *floristic plant geography* and *ecological plant geography*. The former is concerned with a division of the earth's surface into major districts characterized by particular plant or taxonomic groups, whereas the latter seeks to ascertain the distribution of plant communities in relation to their habitats. The floristic approach is very much concerned with the distribution of plant species and their composition in a community or stand.

Biocoenose and *biogeocoenose* are frequently used terms. Biocoenose is not only a synonym for a plant community of any rank, but it is also free from the inference of including the habitat. In reconnaissance studies, methods of combining the biocoenose with its *biotope* or *ecotope* may not be favored, because detailed knowledge of the habitat is either unknown or can be only crudely implied. The equilibrium system formed by combining the biocoenose and biotope has been termed the biogeocoenose (Sukachev in Poore, 1963), *holocoenose* (Friedrichs in Du Rietz, 1936), and the *ecosystem* (Tansley, 1935). Those biocoenoses of highest rank were termed *plant panformations* by Du Rietz (1936) during his field studies; and may be segregated into biocoenoses of somewhat lower rank, which are commonly termed *plant formations*. These in turn may be subdivided into *plant subformations* (sec Chapter 4).

The term biogeocoenose has assumed importance particularly in Russian and German literature. This may be partly due to the restriction of the term ecology to autecology. Biogeocoenose is derived from the Greek roots *bio*—life; *geo*—earth, and *koinos*—common, and it has been defined (Sukachev and Dylis, 1966) as a combination on a specific area of the earth's surface of homogeneous natural phenomena (atmosphere, mineral strata, vegetable animal and microbic life, soils, and water conditions) possessing its own specific type of interaction of these components and a definite type of interchange of their matter and energy among themselves and with other natural phenomena, and representing an internally-contradictory dialectical unity being in constant movement and development.

The biogeocoenose can be equated approximately with Tansley's ecosystem (1935, 1953), defined as a 'unit of vegetation ... which ... includes not only the plants of which it is composed, but the animals habitually associated with them, and also all the physical and chemical components of the immediate environment or habitat which together form a recognizable self-contained entity.' This therefore emphasizes the combination of the biotic community and its environment.

1.3   SCOPE OF PHYTOGEOMORPHOLOGY

The scope of phytogeomorphology can be considered conceptually, spatially, sequentially, or temporally. Broadly, it is intended to infer simultaneous or successive study of vegetation and of landforms, their relationships and their interdependence in the context of the terms previously considered, notably those of *biogeocoenology* and *ecosystem*. However, it is obviously much narrower in concept.

Emphasis is placed on the interdependence of landforms and the higher forms of plant life and not on all environmental phenomena. This is because of the relative ease with which each can be observed, the importance of their relationship to the use and development of land, their amenability to study with the aid of new resources survey methods, and their importance in the planning and management of the rural environment.

There is a natural range of scales at which phytogeomorphic relationships are most clearly expressed. Such scales are hard to define numerically, but are implicit in all landscapes and have been most amenable to and used in mapping. In studying and classifying the landscape and planning its management at the smallest scales, macroclimatic data assume increasing importance over phytogeomorphic data, whereas at intermediate and large scales a phytogeomorphic approach in combination with remotely sensed data may be the most expeditious. At intermediate scales, phytogeomorphic survey can be more cost effective in land use planning than soil survey, but at the largest scales greatest reliance is placed on soil surveys. Frequently soil data will be collected in the field along with the phytogeomorphic, and can then be used as criteria for test-

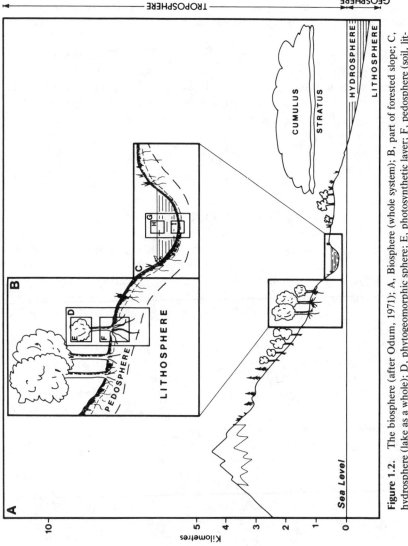

Figure 1.2. The biosphere (after Odum, 1971); A, Biosphere (whole system); B, part of forested slope; C, hydrosphere (lake as a whole); D, phytogeomorphic sphere; E, photosynthetic layer; F, pedosphere (soil, litter, and animal layer); G, lake autecological unit; H, euphotic layer (photosynthetic); and I, mud and hypolimnetic layer.

8

ing, at intermediate scales, the efficiency of landscape classification based on phytogeomorphic data.

Phytogeomorphic mapping ranges from the broad zoning of the earth's continental masses in combination with the macroclimatic data at scales smaller than $1:10^7$ at one extreme, to the articulation, and at scales of $1:10^3$ or larger, of the small variations of landform in a single field or part of a field and with increasing emphasis on soil data at the other. This can be illustrated by considering two extreme examples. The Himalayas are the world's largest mountain mass and have yielded one of its most extensive detrital plains—the Indo–Gangetic. Both mountains and plain and their plant panformations can be marked on the smaller scale maps (e.g., at $1:1$ million). Their relationships approach the limit at which scalar geomorphic and plant distinctions give way to broad-scale climatic and geophysical zonations. At the other extreme, termite mounds in the east African tropics with their own distinctive natural vegetation and soils and with zones of detritus of a few meters around them represent about the smallest objects which can be considered as phytogeomorphic units for agricultural land use. When landform–plant complexes have a smaller surface extent than this, the ecological factors cease to be clearly related to variations of the terrain surface and are far more amenable to analysis in terms of soil series and floristic communities.

Finally, a brief comment should be made on phytogeomorphology within the vertical dimension of the earth and its environment. Phytogeomorphology is primarily confined to the biosphere in which living organisms survive. This, as indicated in Figure 1.2, includes the pedosphere, the hydrosphere, and the lowest part of the troposphere, but not deep into the lithosphere or the stratosphere. In practice, this includes the atmospheric layer up to the tops of the tallest trees and extends below ground to the limits of root development—a total vertical profile that does not exceed 200 m and may be much less than 1 m.

TWO

THE CLIMATIC
SUBDIVISION
OF THE EARTH

2.1 INTRODUCTION

In approaching the relationship between plants and landforms it is important to begin with a consideration of climate. This is because the climate permits a first subdivision of the earth's surface into natural vegetation zones within which phytogeomorphic relationships are readily observed and recorded. Also, climate acts on the earth's surface to mold the landforms and provides the heat and moisture regimes that govern the distribution of plant species.

Ever since the launching of the first environmental satellite in 1960 we have become increasingly familiar with the way regional climate influences the earth's appearance. On color imagery obtained from outer space, the earth appears as a circle whose surface is divided between the blue of the oceans and the greens and browns of the continents, and, within the latter, vegetal zones can be observed in different colors on the earth's surface. In Figure 2.1a (see color insert) the equatorial high rainfall zone appears distinctly among the abundant cloud. On each side of this the cloud becomes sparse and the color changes, and finally the arid desert zone can be observed. Poleward from the desert zone, a similar color sequence occurs in reverse. In Figure 2.1b (see

10

color insert), the water vapor can be seen to thicken into a series of whorls that are anticlockwise in the northern hemisphere and clockwise in the southern. These are only appreciably thin in the low-latitude, high-pressure zones that give deserts on land and allow little vapor condensation over the oceans. This is a simple graphic indication of the importance of climate in zoning the earth's surface. It is confirmed in broad terms when we compare a world climatic map with those of plant formations or land use. Each climatic zone is reflected in the pattern of biotic responses. These are mainly due to combinations of temperature and rainfall and their seasonal variations, with each zone representing a broad phytogeomorphic entity.

2.2 DYNAMIC PROCESSES

The past century has witnessed the introduction and evolution of ideas that have successively brought greater precision and objectivity to climatology and through this to climatic classification. Recent advances have been largely in synoptic concepts, instrumentation, and satellite remote sensing.

An important contribution has been the improved understanding of energy flows and their role in controlling water budgets on the earth's surface. More recently, important developments have been due to the use of satellites for remote sensing of the atmosphere. In addition to providing broad synoptic views, these have in the past two decades yielded regular, long, and effective records of cloud and temperature variations and the distribution of rainfall. With the advances in computation and data processing methods, this has meant significant advances in our understanding of atmospheric processes. In fact, the quantity of data generated from these sources has now become so great that often the main problem has become one of archiving, retrieval, and interpretation rather than one of acquisition (Barrett, 1974).

The incoming radiation from the sun plays an important part in determining the structure of the atmosphere. Its distribution over the earth's surface varies greatly with latitude (solar obliquity) and cloud cover. Although clear skies will permit as much as 80% of the total solar radiation to reach the earth's surface, clouds will reflect 30–60% and absorb 5–20% of it so that well under 50% may reach the ground, and some of this is intercepted by the vegetation canopy. The percentage of incident radiation at the ground surface that is reflected back is known as the *albedo* and varies with changes in the density and the physiognomy of the plant cover and with the reflectance characteristics of the exposed soil and bedrock.

If the earth were not rotating, the excess of net radiation (the balance between all-wave length incoming and outgoing radiation) in intertropical latitudes would simply cause the atmospheric gases to rise and move polewards to offset the net radiation deficit in middle and higher latitudes and the cycle would be completed by a low-level equatorward movement of surface air. However, the air is deflected westward in relation to the increasingly rapid angular momentum it encounters and as it approaches the equator by a vorticity force that gives rise to a ground-level easterly wind and an upper airflow with a strong westerly component. This gives the dominant zonal wind circulation patterns, as shown in Figure 2.2.

Early studies attempted to explain the pressure distribution over the globe. Rising equatorial air would produce a zone of rather low pressure and weak and uncertain winds, traditionally known as the *doldrums*. Poleward of this region the descent of air at about 30° would give rise to the high pressures of the subtropical (Horse) latitudes from which winds would divert both back toward the equatorial low and poleward toward the circumpolar lows that fringe the high-pressure zones of the Arctic and Antarctic.

As early as 1735 Hadley suggested that the tropical circulation was a simple thermal cell of this sort. Ferrel in 1856 postulated a matching indirect circulation in the middle latitudes. It was later assumed that a thermal high-pressure polar cell also existed. Much later Rossby (1949) proposed modifications to this three cell concept to allow for the lack of evidence of the middle indirect

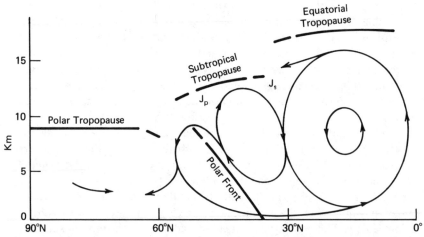

Figure 2.2. Vertical meridional (N–S) plane between equator and pole, showing mean atmospheric circulation. J_p marks the position of the polar front jet stream in the upper westerlies (blowing straight into the page as the diagram is drawn) and J_s marks the very strong westerlies in the subtropical jet stream (also blowing into the page). The subtropical jet stream represents the extra W–E motion relative to the rotating earth at latitude 30° possessed by air that has come (with the vertical and meridional circulation shown) from near the equator (from Lamb, 1972, reproduced by permission of Royal Meteorological Society, London).

cell. He showed that there was a greater exchange of air between middle latitudes and polar areas, mainly caused by large-scale eddies taking the form of frontal depressions. Recent advances, particularly in understanding of the jet streams, have necessitated further modifications. In 1951 Palmen proposed a revised model that was updated by the Royal Meteorological Society (Lamb, 1972); see Figure 2.2. The emphasis today is placed on the spasmodic alternation of zonal (east–west) and meriodinal (north–south) air flows in the westerlies known as the *index cycle*.

Climatic variations at an intermediate scale are known to be caused by the global distribution of the continents and of the uplands and lowlands within them. Thus the lower thermal inertia of the land causes it to heat up in the spring more quickly than the sea, attracting inflowing winds to fill the low-pressure vacuum—a trend that is reversed in autumn. High ground poses barriers to wind circulations so that, for instance, westerly winds that penetrate far into the interior of Europe are excluded from comparable penetration into North America by the barrier of the Rocky Mountains; and the southern hemisphere trade winds are prevented from bringing rain to the eastern part of the Australian interior by the Great Dividing Range.

Warm and cold ocean currents respectively heat and cool the winds that traverse them on their way to neighboring lands. One of the reasons for the extreme dryness of the north coast of Chile is the cooling of the air over the Humboldt current that makes it stable and unable to carry much moisture.

On a smaller scale, diurnal alternations of solar heating and cooling cause offshore and onshore winds. Hills act as barriers causing orographic rainfall on windward slopes and different solar heating on different aspects affecting local winds.

2.3 PAST AND PRESENT CLIMATES

Landscapes have been formed not only by present climates but also by those that have existed in the past. Different climates give rise to different geomorphological processes, soils, and vegetation, and form the zonal and smaller scale variations of the earth's surface that we have been considering. But all areas have experienced profound changes, which may have recurred in cyclic fashion. Variations have been so great that many places in both the northern and southern hemispheres have experienced all types of climate from equatorial to polar at some time in their geological history. Most important for the phytogeomorphologist has been the series of alternations between glacial and temperate climates in middle latitudes and between pluvial and arid climates in the tropics during Pleistocene and Holocene time. We must therefore view landforms and vegetation in terms not only of present climates but also of climatogenesis.

Climatic changes in recent geological history can be considered as essentially due to the periodic expansion and contraction of the high-pressure zone in the polar latitudes. Its expansion caused the glacial ages, each of which included oscillations between more or less extreme cold conditions. These periods are thought to have been associated with the equatorward compression of the other climatic zones, so that westerly depressions in the northern hemisphere tended to take a more southerly path, for instance, across the Mediterranean and North Africa, and desert conditions were moved toward the humid tropics. Interglacial periods witnessed a return to conditions more similar to those of today.

2.4 CLIMATIC CLASSIFICATION

Attempts to classify climate originate from the ancient Greeks who divided the earth into three temperature zones according to the sun's elevation: torrid, temperate, and frigid. Modern zonal classification of climate began only in the late nineteenth century with Supan (1896) who used temperature instead of the sun's elevation. This was followed by Köppen (1900) who combined mean annual and monthly temperatures and precipitation ($P = 2T + 14$) to divide the earth into five climatic zones: tropical humid, dry, warm and cool humid, temperate, and polar. Since then a number of classifications have been based on Köppen's mathematical system.

Vahl (quoted by Bucknell, 1952) followed this system, but based his scheme more distinctly on vegetation distributions; but Thornthwaite (1948), for example, sought to be more fundamental by combining the roles of energy and water into a water budget. He made this the basis for his 1948 classification and his later revisions (Thornthwaite and Mather, 1955, 1957). His complete classification is based on four elements essentially representing the overall amounts of moisture and heat and their seasonal variations. A somewhat similar approach has been followed by Budyko (1956) who related the net radiation available to evaporate water vapor from a wet surface to that required to evaporate the mean annual precipitation and to the latent heat of vaporization. As with the Thornthwaite scheme, this gives numerical values that enable stations to be graded on a scale of aridity. But it goes further in also relating the values to numerical assessments of the runoff coefficient. A more genetic climatic classification based on wind systems was proposed by Hettner (1930).

Parallel developments have taken place in schemes that have sought to zone the earth climatically but in a way that is specifically relevant to vegetation or geomorphology. Holdridge (1947) published with special references to Latin

America a system for classifying natural life zones of vegetation based on the annual precipitation, mean annual temperature, and evapotranspiration (Figure 4.2).

Opinions have differed on whether the classification of climate should be based entirely on climatic criteria or whether climatic classification should be influenced by the vegetation. Gaussen (1955a), for example, considered that it was essential to link vegetation and climate as he did in his vegetation maps, because vegetation is a most effective indicator of current environmental conditions when linked to climatic boundaries.

The term *climatic geomorphology* was introduced by de Martonne and the concept of zones was developed in France by Gaussen who assigned numerical formulas to them (1955b). As mentioned, others have taken the line that climatic classifications, although they may be recognized first in terms of vegetal types or related to landform divisions, should nevertheless be defined solely on climatic criteria.

The original geomorphic approach was to recognize landform regions as a preliminary to analyzing the causative mechanisms. The main pioneer of this approach was W. M. Davis with his distinction between fluvial, glacial, arid, and glacial cycles of erosion. In an attempt to devise a model for explaining the genetic mechanism of such regions, Peltier (1950, 1962) suggested schemes relating given combinations of temperature and rainfall to geomorphic attributes, such as chemical weathering, mass movement, fluvial erosion, and synthesized these into conceptual regional trends.

2.5 SUBDIVISION OF MAJOR CLIMATIC ZONES

The subdivision of the world's major climatic zones is based on more local manifestations of the same causative factors. These can be broadly categorized as the effects of (a) latitude, (b) altitude, and (c) distance from the sea.

The latitudinal zones previously considered are broad bands within which north–south variation is continuous. This variation is in both temperature and in moisture. In the tropics the important latitudinal variation is pluvial and decreases from the continuously wet equatorial climate with small rainfall maxima in spring and autumn, first to the zones of semideciduous forest and deciduous woodland, and then to the Sahel zones with decreasing rainfall more and more concentrated into a short midsummer peak. Poleward from the deserts, through the steppe and then through the temperate deciduous and coniferous forest zones, decreasing temperature and an increasing contrast between summer and winter become important and are accompanied by a steady increase in precipitation. The second marked change of trend occurs at the circumpolar low pressure zone around latitudes 55–60°. Beyond these, the sequence of taiga, tundra, and arctic conditions continue the trend toward decreasing temperatures and stronger contrasts, but precipitation decreases to aridity at the poles.

The change of climatic conditions with increasing altitude is in some respects similar to that due to increasing latitude in the temperate zone. Average temperatures drop by about 0.5°C/100 m. Rainfall increases to a certain altitude, above which it decreases again. The altitude at which it is at a maximum diminishes with latitude from over 1500 m in the tropics to 3000–3500 m in alpine zones, due to the effect of lower temperatures on condensation. There are, however, differences between altitudinal and latitudinal variations. Air thins with increasing altitude, causing its transparency to radiation to increase, with solar radiation approaching the maximum amount possible (the solar constant). The ground is therefore subject to much greater thermal effects where cloud cover is absent. Mountains usually contribute to the formation of clouds and mist and when this occurs, thermal effects are reduced. The mountains may therefore have an exaggerating or a dampening effect on surface temperature variations, depending on the amount and continuity of cloud and mist, and this may result in distinctive habitats for the development of plant communities and the practice of agriculture (e.g., the mist belt in Natal, South Africa).

The third major modification is due to the effect of the distribution of land and sea in inducing the contrast between oceanicity on coastal areas and increasing continentality with distance from the sea. Oceanic conditions are cloudier, wetter, and more exposed to the daily alternation of onshore and offshore winds. Diurnal and seasonal temperature fluctuations are reduced and, on the whole, hot climates are cooled and cold climates warmed, and maximum and minimum temperatures are delayed. In the temperate zone, the greater oceanicity of the western than of the eastern sides of the continents, because of their exposure to westerly air streams and poleward-moving warmer ocean currents, has the effect of "stretching" the transects so that the annual temperature difference between, say, northern Norway and southern Morocco, which are 2500 km apart, is about the same as that between Newfoundland and Florida, which have only half the distance separating them.

The factors causing climatic variations combine to give the characteristic climates of individual localities, which recur in different parts of the world. The middle latitude combination of westerly exposure, winter rainfall, and summer drought, for instance, gives the Mediterranean conditions that may be recognized not only in the type area but also in parts of California, Chile, South Africa, and southwestern Australia where conditions are somewhat analogous.

Schemes have also been devised at national and regional scales to describe climates in terms relevant to agricultural uses. The agricultural climate of

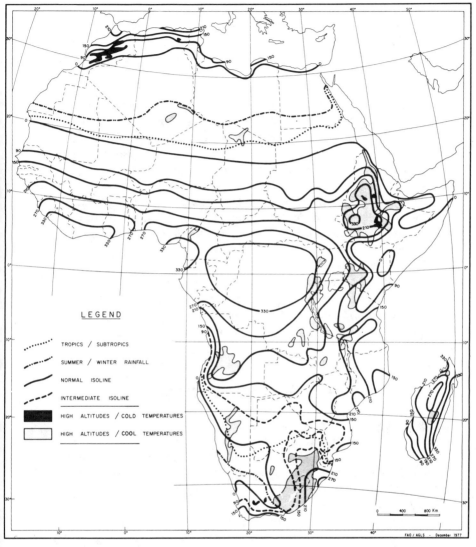

Figure 2.3. Africa: major climatic zones and lengths of crop-growing periods in days (FAO, 1978).

England and Wales has been assessed and mapped by the Ministry of Agriculture, Fisheries and Food, Great Britain (1976), and by Bendelow and Hartnup (1980). The former includes maps of the 80 agroclimatological areas, assessing each in terms of the length and temperature of the growing season, available moisture, and wind exposure.

At a continental scale FAO (1978) has assessed climatic factors in relation to crop yield in Africa. Rainfall reaches a maximum in the equatorial zone and on coasts exposed to the trade winds, and falls to a minimum in the high-pressure subtropical areas, especially where these have sheltering mountains to windward. Altitude causes an increase in rainfall and a decrease in temperature, but almost nowhere in the continent is crop growth restricted by cold as much as by drought.

These considerations led to the recognition of *agro-ecological zones* that differed in the total number of days during the year which usefully combined both (a) rainfall continuously in excess of evapotranspiration and (b) temperatures above a certain minimum level. These zones were "contoured" as shown on Figure 2.3.

The map of agro-ecological zones can be overlaid on the *Soil Map of the World* (FAO-UNESCO, 1974), which shows a number of critical geomorphic and soil attributes limiting to agriculture. The combination of the two maps, when related to particular crops, allowed broad estimates of yield potential at specified levels of management.

2.6   MESOCLIMATES AND MICROCLIMATES

Finally, comment is needed on meso- and microclimates before proceeding to a consideration of landforms and vegetation. Vegetation in particular is influenced not only by the zonal and regional climatic components but also by the local-scale variations known as mesoclimate and the small-scale variations known as microclimate. These are characterized by the greater importance of ground factors.

The most important mesoclimatic factors are aspect, exposure (including effect of wind), surface materials, and vegetation. Insolation occurs on any surface, but increases with its degree of normality to the sun's rays, for example, with a southerly aspect in the northern hemisphere. It reaches a maximum when the equatorward inclination is equal to the latitude (41° in New York and 52° in London).

Daily insolation also causes local winds. During the daytime, air warmed by contact with the ground, especially on the sunny side of a valley, may flow upward as an *anabatic* or *mountain* wind. At night, rivers of cold air, or *katabatic* winds drain downhill and along valley bottoms. In low places the air may stagnate and become exceptionally cold to form *frost hollows* that greatly influence the composition of plant communities.

Exposure is the accessibility of the land surface to climatic effects, notably winds. Practically, it is a function of aspect and altitude. In temperate countries, the exposure of sites becomes more favorable to the extent that they face the sun and are sheltered from strong and cold winds. This particularly influences the distribution of plant species where major differences in relief occur in the north–south direction. In mountainous areas there is a well-known contrast between the south-facing sides and the north-facing sides of valleys; the former in the northern hemisphere having higher snow lines, a longer growing season, and more favorable conditions for agricultural crops and human settlement. In the wine-growing districts of central France and Switzerland, for example, vineyards are located on south-facing slopes below the exposed plateau tops and above the frost-liable valleys.

The effects of microclimate on different surface materials and vegetal structures are dependent to some extent on their different albedos because these determine the amounts of solar heat they absorb. In general, the darker the color of the soil or rock, and the more complete and layered the plant canopy the lower will be its albedo, the more heat it will absorb, and the greater will be its thermal inertia. Albedo values range between extremes of 96% for some new snow surfaces and under 3% for water; but the albedos of most soils and vegetation lie between 10 and 40%.

Vegetation interposes a barrier between the atmosphere and the ground surface, which increases the diffusion of both the inward and outward reflected solar radiations and intercepts a proportion of the rainfall. The leaf area of forest and ungrazed pasture in a humid climate may be several times that of the ground area occupied by the vegetation. On the whole, vegetation, and especially forest, reduces wind speeds and has an oceanic effect on the microclimate in narrowing temperature ranges, delaying extremes, and increasing atmospheric humidity.

THREE

THE GEOMORPHIC FACTOR

3.1 INTRODUCTION

There are three major reasons for proceeding initially with geomorphology, the product of climate acting on geology, and not with vegetation. First, practical studies have shown that any division of the landscape into small and relatively homogeneous units is easier when based initially on landforms rather than on vegetation types. Second, geomorphology describes those more fundamental and enduring features of the landscape that retain an essentially constant character over long periods and are not radically altered by the impact of short-term factors such as fire and drought. Moisture and temperature conditions, for example, change from day to day in response to the weather, and when persisting as a temporary climatic change produce catastrophic changes in the vegetation, which recovers slowly. Herbs and grasses live usually only a few months, or at most a few years, shrubs a few years to a century, and trees seldom longer than several centuries, so that the overall life cycle of plants in any given area is relatively short. Even areas of climax vegetation are subject to periodic change as a result of fire, meteorological catastrophes, and the activities of the population. Soils are more enduring, but they normally reach maturity in periods between a few hundred and one or two thousand years, and may be changed by alterations in the vegetation cover, erosion, and the human activities (e.g., ploughing).

However, under high tropical rainfall (e.g., in Hawaii) recognizable soil properties have been noted to develop in a few decades. Third, geomorphology is causative over larger areas than most other environmental factors. The slope and character of the regolith influence water movements and the resulting vegetation types. These in turn influence the soil's formation.

Although the processes are the temporal expression of the geomorphic factor, the landforms are its physical expression and derive their importance from their enduring nature and from the fact that they strongly influence other factors. A practical appraisal of the geomorphic factor must emphasize those aspects of the environment, such as the relation between surface form and both water availability and soil conditions, which immediately govern the plant habitat.

Terminology in the following sections will be introduced by briefly considering first the geological basis to the tectonic and lithological framework of the landscape, then the modifying surface processes that mold it, and finally the development of a genetic classification of landforms whose relation to vegetation, considered in Chapter 4, forms the basis of phytogeomorphology.

3.2   GEOLOGICAL BASIS

The earth's outer shell or crust is known as the *lithosphere*. It is generally regarded as consisting of two contrasting horizontal zones: an upper zone of relatively light rocks dominated by silica and alumina called the *sial*, and a lower zone of denser rocks dominated by silica and magnesia and called the *sima*. The sial forms mainly the continental masses and varies in thickness from 10 to 50 km and is deepest under mountains. The sima underlies this and rises to the surface immediately beneath the ocean floors. The two are separated by a surface of marked discontinuity known as the *moho* (Figure 3.1).

Tectonic activity occurs mainly within the sial and is the dominant influence in forming the surface of the continents. It deforms rocks and brings them to the surface in a great variety of situations and structural relations, to which their surface forms give expression.

Rocks are aggregates of minerals and differ from one another in form and composition depending on the relative abundance, size, shape, and mutual associations of the grains present. The varieties of rocks are almost endless, but if classified according to mode of origin, they fall into three major classes: *igneous, sedimentary*, and *metamorphic*. Igneous rocks can be either intrusive or extrusive. The former occur as large masses such as laccoliths, lopoliths, or batholiths, as cylindrical plugs, or as dikes or sills that have solidified along fractures or bedding planes, respectively, in other rocks. Extrusive rocks are due to paroxysmal eruptions of magma and include volcanoes, lava flows, and

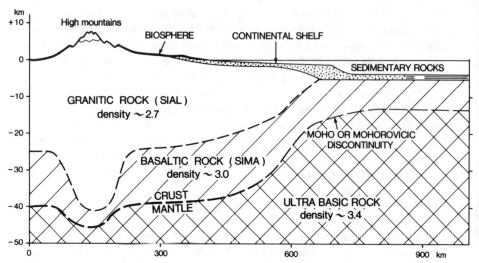

Figure 3.1. Geological profile of the earth's surface (After Holmes, 1956, and Strahler, 1969).

beds of solid ejecta that form irregular layers of solidification at the surface (Figure 3.2).

Sedimentary rocks are composed of bedded deposits left by ice, water, or wind either on land or more usually on the sea floor, which were subsequently uplifted and consolidated. Metamorphic rocks are igneous or sedimentary rocks altered by the pressure and temperatures that accompany mountain building, and are generally harder and more contorted than the original types.

Within these basic groups, the most significant variations are due to lithological differences, especially where specific rock types are selectively concentrated into given areas by sedimentation or evaporation. The most distinctive types of sedimentary lithology are the *siliceous, calcareous, aluminosilicate, evaporite*, and those formed by secondary accumulations that harden on the surface of other rocks—*the duricrusts.*

Siliceous or quartzitic sedimentary rocks are mainly composed of silicon dioxide. They have generally been deposited by relatively fast-moving water or wind. They range in particle size from coarse conglomerates to fine siltstones and malmstones, in hardness from hard quartzites to soft crumbling sandrock, and in purity from homogeneous dune sands to arkoses and greywackes with a large proportion of finer particles. Siliceous materials are insoluble and readily cemented, but develop a relatively evenly spaced rectilinear pattern of fractures along which most water movement occurs.

Calcareous rocks, dominantly formed of calcium carbonate, include marine deposits such as limestones, dolomite, and chalk formed in relatively still shal-

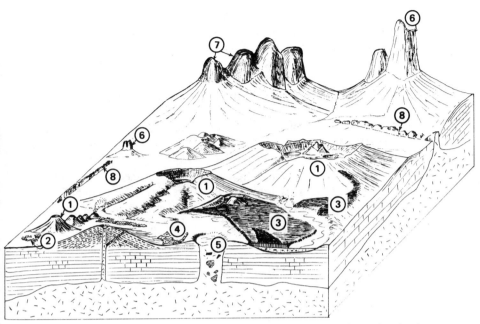

Figure 3.2. Geomorphic block diagram showing common volcanic landforms: 1, volcanic cone; 2, cinder field; 3, lava flow with pahoehoe surface; 4, aa surface; 5, neck; 6, plug; 7, plug dome; 8, dike ridges.

low water, secondary accumulations such as tufa and travertine formed by evaporation from solution under subaerial conditions, and metamorphosed calcite in the form of marble. These may be relatively pure or contain varying amounts of other materials. Marl contains an admixture of clay.

The third major group is composed mainly of aluminosilicates. These are characteristically deposited either from slow-moving water as clay or from ice as boulder clays or tillites that contain coarser particles. The weight of overlying deposits subsequently compresses and solidifies them first into mudstone and then into shales that retain their sedimentary character unless further metamorphosed into slates, schists, or phyllites.

Somewhat similar to carbonates are the evaporite rocks, the most widespread of which are *gypsum, anhydrite*, and *halite*. These are formed by the evaporation of mineralized water in closed basins. In places, for instance, in southwestern Iran and neighboring parts of Africa, subterranean pressures force saline beds up to the surface to form salt domes.

In many areas, but especially under tropical and subtropical climates, soil horizons on sedimentary material can be impregnated with chemicals deposited from charged surface and groundwater that subsequently harden to form indurated surface cappings known as the *duricrusts*. These are classified by the dominant element or compound involved in their formation, that is, *ferricrete* (iron), *calcrete* (calcium carbonate), and *silcrete* (silica). Each is formed where an adequate supply of the dominant element occurs under conditions where it can be mobilized chemically, moved, concentrated, and deposited irreversibly. All types are favored by high temperatures and so tend to be commonest in the tropics.

Ice and permafrost resemble duricrusts by contributing a hardening component to the lithosphere. Ice exists as a surface sheet or as fillings for pores and cracks in the rock. In addition to the polar ice caps, large parts of the globe, especially in the circumpolar lands of the northern hemisphere, have a permafrost subsoil; above this there is a zone of annual freezing and thawing. Gorshkov and Yakushova (1967) estimate that ice sheets and permafrost areas together comprise 20–25% of the entire land area of the globe.

3.3 MODIFYING SURFACE PROCESSES

3.3.1 Weathering

Rocks exposed at the earth's surface are subjected to processes that continually modify their form by removing loose surface fragments and redepositing them elsewhere. Vegetation contributes to the process mainly by root penetration that forms channels facilitating water infiltration by providing a protective cover and by producing organic residues. Those parts of the regolith that have been transported are called *drift*. The processes involved are weathering, transport, and sedimentation.

Weathering is the breakdown and alteration of surface materials into products that are more in equilibrium with the newly imposed ambient physicochemical environment. Most of the large variety of rocks and minerals of the earth's surface do not occur widely and it is possible for practical purposes to confine consideration of the major weathering processes to their effect on the main rock types previously considered.

Physical weathering is the breakdown of rocks by mechanical action leading to their fragmentation and disintegration. It may be caused by either internal or external stresses and takes the form of sheeting, spalling, abrasion, or collapse through undersapping. Sheeting is mainly associated with massive and poorly jointed rocks and results from internal fracturing along planes parallel to the land surface. It is thought to be consequent upon the release of pressures when superincumbent rocks are removed (Ollier, 1969). On a smaller scale, where conditions do not favor sheeting, rocks disintegrate through spalling. Fragments of various sizes are detached by the widening and extension of cracks through the swelling, freezing, and thawing of water, the crystallization of salts, or the penetration of roots.

Temperature change may also cause disintegration of rock matrices through expansion and contraction, especially where juxtaposed minerals have different specific heats and coefficients of expansion and are exposed to extreme temperature variations, as in deserts or under the influence of fire. Rates of disintegration are greatly increased where water is present and able to penetrate into the interstices of the rock. Rocks may also be worn away by the mechanical abrasion of particles moving over the surface under the impulsion of ice, water, or wind.

Chemical weathering in a water medium attacks exposed rock faces, the extent of attack varying with the rate of water movement and the solubility of the rock. Weathering is greatest in halite, gypsum, and limestone. Solutions may precipitate chemicals in the interstices of rocks that can cause volume changes and thus also increase physical weathering. *Oxidation* is due to the abundance and the reactivity of oxygen in soil water. Its main practical effect is to attack the ferromagnesian minerals, raising the degree of ionization of the included iron and manganese. Change in the former is often recognizable from the appearance of the red and yellow colors of hematite and limonite, respectively. *Reduction* is the reverse process, which occurs when oxygen is excluded as a result of waterlogging. Anaerobic bacteria extract oxygen from the rock materials by reducing sulfate, nitrate, carbonate, and other ions. *Hydration* is the absorption of water by minerals, causing swelling. For example, anhydrite changes to gypsum and hematite to limonite.

Rock decomposition is greatly hastened by the presence of organic acids. *Hydrolysis* is the action of ionized hydrogen and hydroxyl ions on minerals, and reflects the degree to which water is ionized. This increases markedly with temperature, almost doubling between 0 and 30°C. The effect is much increased by *carbonation*—the formation of weak carbonic acid from dissolved carbon dioxide, which in saturated water increases the hydrogen ion concentration about 300 times. Hydrolysis and carbonation are thus especially effective in combination. They speed the solution of calcareous rocks and deplete soil silicate clays of their adsorbed bases.

Some living organisms such as bacteria, algae, lichens, and mosses are able to colonize rock surfaces and are important in rock decomposition and preparing soil for subsequent colonization by higher plants and animals.

3.3.2 Transport

Weathering is a slow *in situ* process in landform development, whereas the transportation of materials is much more readily observed. The character and surface forms of drift deposits considered in later sections reflect their genetic processes, notably the distinction between *glacial, colluvial, alluvial, lacustrine*, and *coastal* types.

Ice sheet and valley glaciers are viscous and carry a mixture of materials that they deposit either by forward movement at their peripheries or from sediment-charged meltwater. They are characterized by a lack of stratification and a mixture of particle sizes from the largest stones to small colloids that indicates the origin of the name—boulder clay.

Colluvial material is that which falls mainly from its own weight. Rock slopes are rendered unstable by weathering in zones of weakness behind masses that are poised at an angle exceeding a critical threshold that approximates to the angle of rest of the same materials in unconsolidated form. In humid climates, collapses are commonest where the surface is covered with a mantle of regolith. Downward movement in this mantle occurs where the slope angle exceeds that of the stability of the material. Between gullies this leads to surface creep or deeper mass movements. *Surface creep*, or solifluction, is a downward slide of soil over a more indurated subsoil along a junction usually lubricated with water. Deeper mass movements depend on the moisture state of the regolith, slipping in masses that are relatively dry, slumping when above the plastic limit, and flowing when above the liquid limit.

Alluviation occurs when the sediment is mainly transported by water rather than directly by gravity and is deposited where diminished gradient or added sediment load decreases its carrying capacity. River valleys commonly experience periods of erosional incision alternating with periods of stability. They deepen their valleys and extend them upstream during the incising periods and widen them by sideways excavation in more stable periods, depositing materials laterally and terminally under conditions of slackening flow.

Basin sites with internal drainage in humid regions are generally filled with lakes and marshes that experience sedimentation of very fine particles, especially toward the slack water at their centers. In arid climates they may evaporate to dryness every year, leaving a ground surface impregnated with salts, with those of greatest solubility likewise zoned toward the center.

Although both water and air are ubiquitous and have an entraining and carrying capacity proportional to the cube of their velocities, the latter's combination of greater depth with occasionally high velocity makes it potentially able to create bedforms many times the height of those of river valleys. Its lower density, however, confines its effectiveness to areas where the surface materials are fine, loose, and mobile, that is, mainly to deserts.

Eolian deposition occurs wherever the wind becomes supersaturated through a reduction in velocity by the convergence or interruption of its flow. In general the size of particles transported near the ground tends to fall in the 100–400 μm range, because those smaller have too much mutual adhesion and those larger are too heavy to move far. Because most particles in this range tend to be of quartz, this mineral dominates their composition. Eolian action is also responsible for the deposition of *loess* in China and the Mississippi basin.

Coasts are the product of waves and tides. These exert their greatest force against headlands, acting somewhat like a horizontal saw to form a steep wall called a *marine cliff* and a basal indentation called a *wave-cut notch*, which, when extended by further cliff retreat, becomes an *abrasion platform*. The detached materials, to which are often added estuarine or deltaic sediments derived from rivers, are then moved by the sea to form *beaches*. These are composed of detached materials, comminuted, moved, and sorted by combinations and alternations of coastward *swash*, seaward *backwash*, and longshore drift, caused by advancing, retreating, and laterally moving waves. The details of beach form and composition, which are often highly complex, reflect the sequences, strengths, and directions of these movements.

3.4 LANDFORMS: CLASSIFICATION

Landforms are a reflection of their geological foundations modified by the processes of weathering and transport considered above, that is, they can be viewed as rock–climate–process systems. Because of the complexity of such systems, a number of simplified approaches have been made in classifying the landforms that result from them. As explained in Chapter 2, it is generally necessary to start with a broad classification of landscapes into major climatic zones and then to consider the other governing factors within each.

First, landscape classifications can be *morphometric*, based on a geometric analysis of surface form with relatively little emphasis on rock type or mode of origin. Comprehensive schemes require the inclusion of classes of gradient, aspect, relief amplitude, and the identification of such derived attributes as *stream order, bifurcation ratio, stream length ratio*, and *drainage density*. Stream orders, as defined by Strahler (1964), reflect the importance and nature of their tributaries, ranging from one for fingertip streams to five or six for major trunk rivers. To qualify for the next higher order a stream must unite two tributaries from the next lower order. The bifurcation ratio of a consecutive pair of orders is the total number of streams in the lower order divided by the total number in the higher order. The stream length ratio is the total length of streams in the lower order divided by the total length of those in the next higher order. The drainage density is the average length of stream channel per unit area of basin and is thus indicative of the fineness of erosional texture of the landscape.

Second, landforms may be classified *lithologically*. Different rock types give rise to readily identifiable types of landscape. Quartzites, for instance, are resistant to erosion and normally form higher ground. Granites generate relatively homogeneous but steep mountain slopes that in arid and semiarrid climates often abruptly change to much gentler footslopes. Limestones give rise to *karstic* phenomena of solution and collapse, sometimes causing steep gorges and underground caverns, and shales give landscapes of low undulating relief.

Third, landforms can be considered from the point of view of their origins and classified *morphogenetically* in terms of their mode and sequence of development. The landscape is viewed as a palimpsest in which each recognizable stage of evolution consitutes a separate unit. This type of classification has explanatory value and gives the best approach to the prediction of future changes. Where the emphasis is on the relative ages of the different units, the approach is termed *morphochronological*.

The approach adopted in this chapter attempts to harmonize these approaches and emphasizes those features of the landscape that are significant to genesis, relevant to actual and potential land uses, and readily recognizable.

The largest landforms can be considered to be the major land masses representing the protocontinents of Laurasia and Gondwanaland, and the vast sediment-filled circumvallations between and around them. These have been folded into the vast mountain chains that traverse the world in a belt from Morocco to New Zealand and circumscribe the Pacific Ocean.

Both shields and circumvallations are themselves assemblages of relief features of widely varying size and complexity; their hierarchical subdivisions are considered in Chapters 5 and 6. It is most convenient in categorizing landforms, however, to describe them first at the generally intermediate "natural" scales that are seen by the ground observer and from which the everyday terminology of hills, valleys, plains, and so on, are derived.

The first distinction is between degradational and aggradational forms, the former being composed of consolidated materials mainly undergoing erosion, and the latter composed of the unconsolidated materials derived from their denudation.

3.4.1   Degradational Landforms

As mentioned in Section 3.2, the most significant distinction between rocks in determining landform type is that between those of igneous, metamorphic, and sedimentary origin.

It is convenient to start with volcanic rocks because of their ease of recognition. The most characteristic shape is the *volcanic cone* with a central crater containing one or more vents that may be accompanied by small neighboring adventive cones. The surrounding surface is usually covered by volcanic debris in the form of a *cinder field* if extruded in solid form or a *lava flow* for liquid. Lava flows give two major types of surfaces: smooth and billowy or rough, broken, and piled into chaotic masses of blocks called *pahoehoe* and *aa*, respectively.

Weathered volcanic hills soon develop a rounded form and a differentiation between parts of differing competence in resisting erosion (Figure 3.2). Mount Pisgah, California is an example of a volcano around which different stages of weathering on different ages of flow can be seen. Sometimes the core of the volcano is left exposed as a residual more resistant *neck* or *plug*, or, where rounded, a *plug dome*, from which may radiate upstanding ridges formed by dikes.

Intrusive igneous rocks result from the relatively slow solidification of magma (Figure 3.3). When tectonic forces elevate them or circumdenudation isolates them as uplands, the directions of the original faulting are often preserved in the resulting, often tilted, block mountains, which when juxtaposed form *horsts* and *grabens*, as in the basin and range area of the American West or the Vosges and the Black Forest *massifs* on either side of the Rhine rift valley. Weathering along lines of fracture causes the surface relief to reflect the structure of the parent rock.

The three scales of transformation that go by the name metamorphism are *regional, dynamic*, and *contact*. Regional metamorphism results from deepseated plutonic activity that causes complex patterns of lineation and fracturing of the rocks at continental scale. Over smaller areas dynamic metamorphism associated with folding and faulting alters the structures of rocks by directional pressure. It causes their fragmentation, realignment, and recrystallization, imposing a stress-oriented directionality. In mountainous or shield areas this gives rise to topographic alignments that are related to igneous outcrops caused by the same diastrophic forces. *Contact metamorphism* is the change that is caused by the intrusion of igneous rock against a surrounding mass and its effects are generally confined to the narrow friction zone between them. It gives rise to physical and chemical changes that form new compounds and alter the rock structure. Where such compounds are schistoid or kaolinitic, the junction zone will be softer and more vulnerable to weathering and erosion than where they are siliceous.

Sedimentary rocks (Figure 3.4) are those that have retained the form of the stratified deposits that compose them, although they may have been elevated, folded, tilted or faulted since their deposition or a combination of these. Folding occurs along axes, giving rise to *anticlines* and *synclines*. When the axial plane of a fold is inclined to the vertical, the fold is said to be asymmetrical; when it is so much inclined that one limit of the fold becomes doubled under the other, it is an *overfold*. If the orogenic (mountain-building) push continues and the axial plane approaches horizontality, we have a *recumbent fold*. Examples can be seen in both the Appalachians and the Alps.

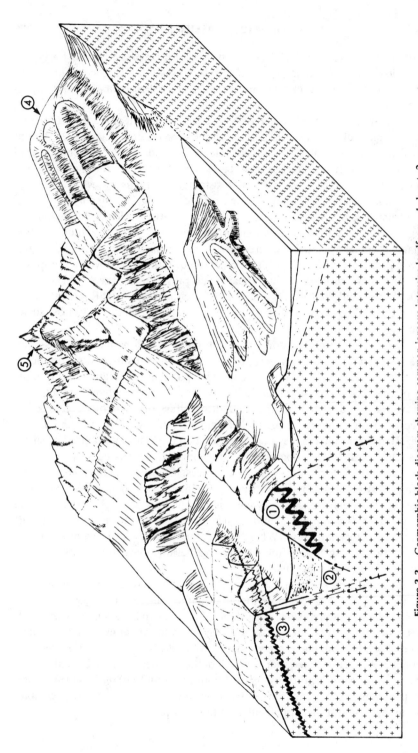

Figure 3.3. Geomorphic block diagram showing common igneous intrusive landforms: 1, horst; 2, graben; 3, tilted block mountain; 4, rounded landscape characteristic of humid climates; 5, glaciated mountains.

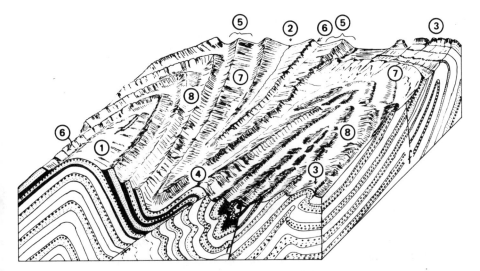

Figure 3.4. Geomorphic block diagram showing common landforms on sedimentary rocks: 1, anticline; 2, syncline; 3, overfold; 4, recumbent fold; 5, cuesta; 6, escarpment; 7, dipslope; 8, strike vale.

Level-lying sedimentary rocks give rise to plains which, when elevated in the landscape, are subjected to dissection in a generally dendritic pattern, that is, resembling in plan view the branching of a tree. This often leaves the higher ground as fragmented tabular remnants whose surface may be formed by a relatively resistant bed that protects those underlying it. The character of the landscape is determined by the elevation of the plateaus and their degree of dissection. Irregularities along escarpment edges tend to be exploited and magnified because reentrants have larger catchments than spurs and so enlarge more quickly, thus capturing more and more of the drainage of the plateau surface and accentuating the contrast in relative advancement between the promontories and embayments that give the plateau edge a frayed appearance, and leaving *outliers* as in the Grand Canyon area or on the dissected hamadas of North Africa.

As the angle of dip of the rocks is increased, the tendency for differentiation of promontories from reentrants is proportionately decreased. When steeply dipping, they form narrow *strike ridges*. When tilted through faulting or folding, sedimentary strata experience erosional differentiation of the competent (resistant) from the incompetent (nonresistant) beds. This leads to the formation of a *scarp-and-vale* topography consisting of asymmetrical ridges, known as *cuestas*, with relatively steep *escarpments* and relatively gentle *dipslopes*, separated by *strike vales*, that give a landscape such as that in the eastern coastal plain of the United States and the lowlands centering on London and Paris. Figure 3.4 illustrates some of these relationships schematically.

Sedimentary areas are thus broadly composed of level or sloping surfaces, cliff faces, footslopes, and valleys or plains. Within these basic structural patterns, the most significant variations are those that reflect the major lithological types of sedimentary rocks such as siliceous, calcareous, aluminosilicate, evaporite, and those that are duricrusted.

Landforms on siliceous and calcareous sedimentary rocks reflect the competence of these materials in resisting erosion (Figure 3.5). For siliceous rocks this tends to be at a maximum where the particles are coarsest and where the binding matrix is hard and insoluble, especially where its surface is cemented. For example, although the loosely cemented and unfissured greensand deposits of the English Weald and the harder, coarser, and more fractured Millstone Grit of the Pennines form eminences above the surrounding shales, the soft undulating landforms of the former contrast strongly with the sharper outlines and tor-crowned cliffs of the latter.

Despite this variety, however, rocks in this group have certain common characteristics that give rise to similarity of form. They are more competent than

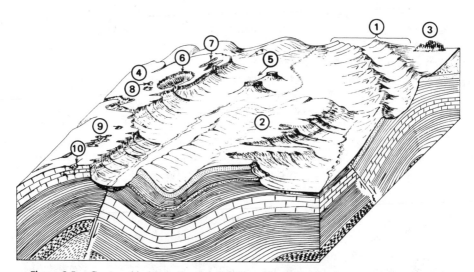

Figure 3.5. Geomorphic block diagram showing common landforms on calcareous and siliceous sediments: 1, sandstone anticline; 2, sandstone syncline; 3, massive castellated sandstone; 4, limestone plateau with 5, tabular outliers and karstic features; 6, amphitheatres; 7, cluse; 8, doline; 9, uvala; 10, sink hole.

finer textured rocks and so generally form the higher ground. Because most water movement in sandstones is along fractures and these fractures tend to be rectilinear, the rock breaks into relatively evenly spaced blocks. Because the rock is insoluble and its structure granular, the mechanical disintegration of its surface yields materials subject to relatively rapid removal by wind and water. The residual rock loses its most vulnerable edges and corners, giving rise to smooth rounded forms that here and there stand up as *tors* or castlelike forms where the matrix is exceptionally strong. Contrast between hard and soft matrices is especially marked in deserts because of the dominance of mechanical over chemical weathering. Where large-scale fracturing is more marked than small scale, hills stand more abruptly above their coarse detrital plains. The dominance of small-scale fracturing gives areas of gentler relief and finer textured detritus. Where wind action has free play, dunes develop.

The distinctive features of landscapes developed on calcareous rocks (Figure 3.5) derive from two main characteristics of the materials—permeability and solubility. Level limestone plains and plateaus erode in a similar manner to those on sandstone except that they are subject to solution phenomena known collectively as *karst* and consisting of surface and subsurface recesses and hollows ranging in size from a few centimeters to several kilometers. The smallest features are the surface roughnesses on the rock caused by differential solution that leave points and edges, some of which are due to contained silica. These are called *lapies* when they are more pronounced and have a pitted or honeycomb form. When the surface is bare it may be fissured either into large flagstones or into a network of ridges and furrows called *clints* and *grykes*, respectively. Larger hollow features in ascending order of size are *solution pits, sink holes*, hollows floored with soil (called *dolines* when small, *poljes* when large, or *uvalas* when two or more have coalesced into one). Underground the same areas are often riddled with caves, grottoes, caverns, and tunnels. The deep gorges characteristic of rivers in limestone country are often a result of underground streams whose roofs have collapsed. The development of these features is at a maximum where the surface has been longest exposed, where the rock is pure and massive, and where rainfall is abundant, especially when the vegetation generates a high concentration of organic acids. In arid areas karstic features are less marked and usually take the form of shallow basins filled with fine sediment (Mitchell and Willimott, 1974). Because of the paucity of quartz or other fragments of appropriate size, dunes are small or absent.

The erosional retreat of escarpments that terminate limestone plateaus and cuestas shows the same patterns as discussed previously except for the greater steepness of slopes and sharpness of cliffs resulting from solution followed by collapse, especially where this takes place at spring lines above subjacent impermeable beds. Slopes are gentler where the beds are less massive or are intercalated with the shale or marl deposits.

In areas of tilted sediments, karst development is reduced because the surface gradients increase runoff at the expense of infiltration. Valleys are characteristically dry except when runoff temporarily exceeds evapotranspiration and the infiltration capacity of the underlying rock, for instance, in winter.

Secondary deposits formed from the evaporation of lime-rich water may sometimes give rise to prominent landforms, especially in arid areas. They are called *mound springs* or *tufa domes* where the water emerges at a point or *travertine dikes* where they are along a fissure.

Because they combine softness and impermeability, clay and shale are the sediments most vulnerable to erosion. They generally form the lowest portions of the landscape—the basins, valley bottoms, and coastal embayments. In ridge-and-valley landscapes formed by the differential erosion of tilted beds of contrasting hardness they tend to form the valleys. Where other rock types are absent the relief is characteristically subdued and in humid climates gently undulating. Even narrow intercalations of other materials may form conspicuous landscape features. Under conditions of restricted vegetation cover and torrential rain, they tend to develop the pattern of fine intricate dissection known as *badlands*.

Evaporite deposits are too vulnerable to solution to survive at the surface in humid climates, but they are widespread in arid areas. Because of their greater solubility, evaporite landscapes show phenomena of caving and collapse analogous to those of karst areas, but their weaker cohesion and limited extent make them less topographically significant.

Duricrusts occur as surface sheets following the contours of the topography on which they are formed. They have greater resistance to erosion than is possessed by the surrounding nonduricrusted areas, sometimes leading to an inversion of relief whereby those areas originally low become prominent. These may survive as physical evidence of formerly wetter conditions in areas where the climate has changed, as in some ferricreted areas (i.e., with a surface crust cemented by iron oxides) in semiarid Australia (Mabbutt, 1977) and in calcreted toposequences of alluvial fans in the Maghrebian pre-Sahara (Joly, 1962).

Permafrost areas are those in which the subsoil is permanently frozen. They are widespread, especially in high latitude areas of North America and Eurasia, and show a distinctive complex of relief forms. The most conspicuous are those that result from the action of melting and freezing—*thermokarst* and *hydrolaccolith* features, respectively.

Thermokarst is formed by the thawing of subsurface ice and the resulting subsidence. It generally takes the form of hollows ranging in width from a few centimeters to several kilometers and in shape from narrow chimneys to broad basins, some of which may hold ponds or lakes. The uneven hummocky surfaces and disorganized drainage patterns bear some resemblance to karst.

A hydrolaccolith is a special form of frost mound resulting from the bulging up of frozen ground under hydrostatic pressure from below where it is caught between downward freezing from above and permafrost or impermeable rocks below. When large, such features are known as *pingos*. On a smaller scale, *cryoturbation* (frost heaving) causes the upheaval and convolution of soil profiles often associated with patterned ground.

3.4.2 Aggradational Landforms

It is difficult to draw a clear distinction between degradational and aggradational landforms. Most sites experience both conditions at different times. Hard rock areas can be buried and recent alluvial deposits can be eroded. Nevertheless, in general, the consolidated landforms hitherto considered are dominantly suffering erosional loss, whereas those on the unconsolidated materials that surround them are mainly experiencing depositional gain. Aggradational landforms reflect the processes that have formed them (Figure 3.6). Those deposited by ice are spread in uneven sheets as *till plains*. These characterize wide areas in northern North America and Europe. They contain *drumlins* and *moraines* that reflect the history of ice movements in relation to the periglacial

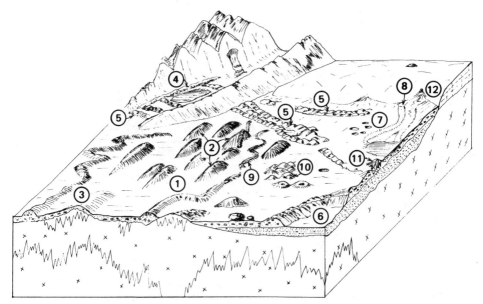

Figure 3.6. Geomorphic block diagram showing common glacial landforms: 1, till plain; 2, drumlins; 3, rockdrumlin; 4, lateral moraine; 5, terminal moraine; 6, sandur; 7, kettled sandur; 8, valley terrace; 9, esker; 10, kames; 11, pingo; 12, meltwater channel.

topography. These deposits vary in surface form but give a generally uneven hill and kettle relief, often tens of meters in amplitude, consisting of mounds, ridges, and basins arranged with a poorly developed drainage pattern. Fluvio-glacial flow gives better stratified and sorted materials that form outwash plains, sometimes known as *sandurs*, and valley terraces. Because melting is often rapid, textures can be coarse. Materials deposited beneath the ice by melt-water may form *eskers*, which have the form of either mounds called *kames* or sinuous ridges called *osar* (*singular, os*).

Colluvial landforms are those formed by material that has mainly been de-posited by falling under its own weight. They surround uplands in all climates (Figure 3.7). They reflect the nature and relief of the rocks from which they are derived, especially differentiating between those that are bare and those that are covered with a soil mantle. Deposits derived from bare rock surfaces reflect the sizes of the rock fragments that compose them, which in turn reflect the fracturing in the original rock. Where macrojointing is more important than microjointing, slab failure occurs and large blocks fall to form rough piles of boulders. Where microjointing is dominant, the smaller spalls drop or ava-lanche into *talus slopes, talus glaciers,* or *dejection cones* depending on their form and degree of lateral confinement. Where the rock disintegrates mainly

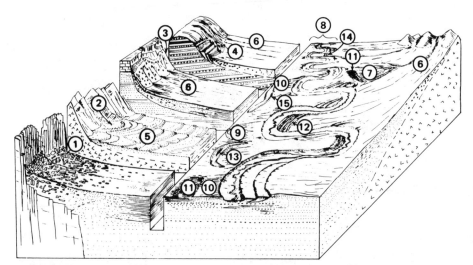

Figure 3.7. Geomorphic block diagram showing common colluvial and alluvial landforms: 1, talus slopes; 2, talus glaciers; 3, dejection cones; 4, terracettes; 5, mudflows; 6, alluvial fans; 7, deltas; 8, active flood plain; 9, cover flood plain; 10, backslope; 11, backswamp; 12, point bars and swales; 13, oxbow lake; 14, cut-off; 15, channel-levee remnant.

into grains at the surface the resulting finer deposit seldom remains as talus because of its lower angle of rest and greater vulnerability to removal.

Colluvial landforms result from surface creep or deeper mass movements. The former can be seen in hillside scars, the piling of earth on the upslope side and exposure of the downslope side of obstacles, and bent poles and tree trunks. The result is a thickening of the regolith at the bottom of a slope at the expense of that at the top. Deeper slides in cohesive material result in the formation of one or a sequence of slumped masses topped by small *terracettes*, and separated by relatively steep back walls or *clifflets*. Material that is partly fluid when it moves tends to solidify into an uneven hummocky surface between a steep exposed back wall and a lobate outer edge. When the materials exceed the liquid limit, they form *mud flows*, and when under especially strong gravitational impulsion, *mud spates*. These present a rough, boulder-strewn, but essentially level surface with lobate edges marking the limits of their advance. A well-known example is the Slumgullion mudflow in Colorado.

Abrupt lessening of slope in river valleys causes the formation of braided channels and alluvial fans, which can be seen especially clearly in arid areas such as the American Southwest. Alluvial fans commonly occur where upland streams debouch onto a plain, and owe their form, extent, and composition to the nature of the contributing catchment. They are usually gentler in slope than gravity deposits, concave in longitudinal section, and marked by anastomosing channels radiating from the apex, with progressively finer soil textures outward from the same point.

The deposits over thousands of years of seasonal and perennial rivers within their valleys normally take the form of terraces and deltas, such as can be seen along such major rivers as the Mississippi, Ganges, Amazon, Yangtse, and Nile. A common pattern is that the river has a history of periods of erosional incision alternating with periods of stability, leading to the development of flights of terraces. In general the highest are the oldest and driest with best developed soil profiles, but show the fewest traces of the original alluvial deposition.

The differentiation of alluvial landforms is most clearly seen in large river valleys. Near the river is an *active flood plain*. This may be on a levee bounded by *backslope* and *backswamp* areas and shows evidences of current meander action such as *point bars* and *swales* on the inner side of meander bends, *undercut slopes* on the outer side, and *cutoffs* where the river has shortened its course. Next to this is a *meander flood plain*, recently abandoned by the river but still subject to occasional inundation. This often shows a scroll pattern from the air with traces of former meander channels locally containing *oxbow* lakes. Farther from the current channel is a *cover flood plain* where the traces of past river action are almost completely effaced. Its surface is seldom flooded and is generally level and featureless. It is here that raised *channel levee remnants* may survive. These represent earlier courses of the river whose survival is due to the greater erosion resistance of the coarser textures laid down in their beds. Deltaic deposits near the mouths of rivers reproduce on a broader scale and with lower relief the braided characteristics of alluvial fans. They are characteristically level swampy areas crossed by distributory channels from the river. Longshore currents along the coast may construct beach bars that impound river waters in inshore marshes and lagoons.

In humid climates basin sites in valley alluvium are generally filled with permanent lakes or marshes, but their margins are dominated by the character the vegetation gives to them and this in turn depends on the depth and base status of the water.

Wind (eolian) action results in the deposition of landforms conspicuously different from those due to water (Figure 3.8). These are conventionally known as dunes, but the more general term of bedform is often used because of its lack of a specific size connotation. In arid areas these have a great variety of sizes and shapes. There is some evidence that there are three preferred orders of magnitude, which can occur in juxtaposition: (a) *ripples*, a few millimeters to a

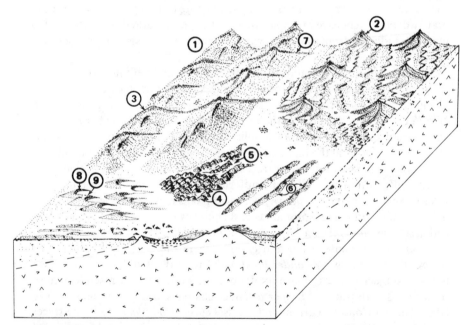

Figure 3.8. Geomorphic block diagram showing common aeolian landforms: 1, draa; 2, oghroud; 3, demka; 4, aklé; 5, ansguie; 6, seif; 7, silk (plural slouk) dunes; 8, barkhan; 9, linguoid dune. Large sand seas in background are ergs.

meter or so high, caused by diurnal wind variations; (b) *dunes*, up to a few tens of meters high, related to cyclonic disturbances; and (c) *draa*, hundreds of meters high, due to zonal wind circulations.

Dunes and ripples in humid regions are frequently found along coasts where the mobile beach material forms a hummocky terrain up to a few meters in height. In deserts, dunes can be longitudinal to the wind such as *seif* and *slouk* dunes, transverse such as *barchane* and *linguoid* dunes, or complex such as *aklè* or *ansguiè* dunes, which represent disorganized and partly aligned "rough seas," respectively. Draa exist only in the large sand seas known as *ergs*. They are either long parallel ridges as in the Rub Al Khali of Arabia or have regularly spaced peaks alternating with bare hollows as in parts of the Algodones dunes of California. The high infiltration capacity of all forms of sand accumulation favors rapid colonization by vegetation when there is any increase in rainfall. Where this is a long-term trend, dunes become fixed as in the sand hills of Nebraska.

Loess deposits differ from sand dunes in being spread out by the wind into a surface mantle whose thickness can vary from minor admixtures in the soil to depths of over 100 m. They occur mainly near the margins of arid or formerly arid regions. Apart from important occurrences in China and in the Parana Basin of South America, they occur in three distinct areas of the United States: (a) capping higher ground in the Mississippi Basin; (b) in Kansas, Nebraska, Iowa, and Missouri derived from the arid regions farther west; and (c) in the Palouse area of Washington and Idaho derived from glacial outwash. Loess-covered landscapes tend to follow the contours of the underlying terrain modified by some wind-formed undulations or valleys, the abruptness and steepness of whose sides are due to the cohesion of the interlocking platy silt-sized particles.

In dry climates, lakes often evaporate to dryness every year and leave various kinds of ground surface depending on the nature of the alluvial material and salts. Studies of these surfaces have been reviewed by Cooke and Warren (1973) and Mabbutt (1977). In general those fed by groundwater tend to have hummocky surfaces, whereas those dominated by surface flooding have smooth, finely cracked, silty surfaces. Many large lakes (e.g., Lake Chad in Africa) produce distinctive peripheral landforming associated with climatic cycles.

The sea builds beaches (Figure 3.9) from eroded coastal detritus and river-mouth alluvium. As the maximum vigor of waves is against headlands, these lose material that is then drifted along the coast into coves and bays. The material is then sorted in a seaward direction to leave the coarsest fragments in the intertidal zone and becomes finer offshore, ending in mud flats that are only exposed at the lowest tides. The sequence may not be continuous and may exhibit a series of parallel strand lines or dunes parallel to the coast. Embankments of sand and gravel accumulate offshore on the sea floor where inshore or longshore waves and currents are arrested, and may grow high enough to appear

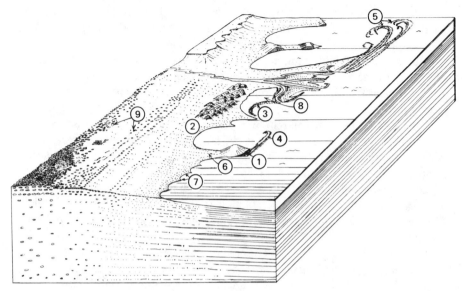

Figure 3.9. Geomorphic block diagram showing common coastal landforms: 1, cliffs; 2, bayhead beach; 3, spit; 4, hook; 5, recurved spit; 6, tombolo; 7, cusps; 8, lagoon; 9, stony beach ridges.

above the surface. Where the longshore component is strong and unidirectional, the material may form *spits*, which are sometimes recurved into *hooks*, or extend out to islands as *tombolos*. Alternating currents from opposite directions lead to the formation of *cusps*. Coastal landforms or their remnants also occur inland from the coast if they have been isolated from its immediate vicinity by tectonic uplift or alluvial deposition from the landward side.

3.5 LANDFORMS: GLOBAL DISTRIBUTION

As indicated in Chapter 2, the world can be divided into climatic zones each of which produces distinct types of landscape. The zonation is in fact more complex than this because the Tertiary and Quaternary climatic alternations have left in many areas the imprint of processes or successions of different processes that are no longer indigenous to the regions in which they occurred. In particular, the equatorward displacement of climatic zones during the Quaternary glaciations has left glacial and periglacial features in the temperate zone, pluvial features in the arid zone, and eolian features in the semiarid tropics. Landforms are thus climatogenetically, as well as climatically, determined, and the complex of differences thus caused can be defined as *climatovariance*. Nevertheless the processes active in each climatic zone today are sufficiently similar to those of the recent geological past for them to have developed broadly recognizable and predictable geomorphological features. There have been two approaches to the classification of zonal landforms. Peltier (1950), followed by Tanner (1961), Leopold, Wolman, and Miller (1964), and others sought to identify regions theoretically by partitioning the space on temperature–rainfall graphs into categories relevant to specific geomorphic processes such as frost action, chemical weathering, mass movement, and wind action. For instance, chemical weathering is intense where both rainfall and temperature are high, moderate where they are intermediate, and weak where they are low. Such analyses have a theoretical and explanatory value, but they are inevitably highly generalized.

On the other hand, Passarge (1926), followed by Büdel (1963), Tricart and Cailleux (1972), and others have sought inductive regionalizations on the basis of the assemblages of landforms in different climates. Tricart and Cailleux also recognized some analogy with the zonal concept in soil science. Landforms with the characteristics of certain climatic zones are called *zonal*. Widespread features common to all climates such as volcanoes, beaches, and bare rock exposures are termed *azonal*. Others that occur outside their normal zones are called *extrazonal*. Examples are *prolongations* such as river valleys that extend the characteristics of catchments in another zone and *survivals* such as outlying hills that reproduce the conditions of a related mountain area from which they are separated. Other landforms that occur in several but not all climatic zones, such as maobile sand dunes or alluvial fans, are known as *polyzonal*.

The zonal landforms of humid climates can be broadly distinguished from those of arid by their less strongly marked topographic angles, smoother slopes, and the prevalence of stream profiles in the relative equilibrium which results from continuity and regularity of flow. The humid tropics are characterized by the absence of frost action and by intense chemical and biotic weathering that form abrupt and steep hillslopes, usually without clear pediments, and lowland plains whose surface materials contain a high proportion of clays. River valleys are characteristically choked with fine sediments that lodge in backwaters, basins, estuaries, and along coasts and are often covered by dense vegetation. This cover, combined with low wind speeds, renders eolian action negligible.

In the semiarid tropics, often referred to as the *sahel*, rivers are more often seasonal, erosional gullying is more widespread, and there is a greater discontinuity between wash slopes and piedmont slopes. Loess is deposited by wind from more arid areas. Sand dunes may form locally, but are not widely developed except where relict from drier palaeoclimates.

In the arid zone, the morphogenetic processes due to vegetation are slight or absent, leading to a dominance of physical over chemical weathering. Runoff is rapid and intense, but because of the lower rainfall there is less gullying and erosion than in the semiarid tropics. Rocks are barer and sharper. With increasing aridity, wind replaces running water as the dominant surface force and the landscape is characterized by deflated stony hamadas, scoured rock surfaces, and mobile dunes. The seasonally intense heat brings high evaporation, causing basin sites to become salinized. Several years may elapse between rainstorms.

In the zone of steppe grassland on the poleward margin of the deserts there is a return to the semiarid landforms previously mentioned, but with certain differences. The cooler temperatures make for lower evaporation and hence, especially where there is a winter rainfall maximum, a tendency for more runoff and erosion with the same total rainfall. On the other hand, chemical weathering is less intense so that, for instance, there is less occurrence of the montmorillonite clays that form vertisols on low ground in the tropics.

The humid temperate zone bears some resemblance to the humid tropics, but the landforms are markedly different. The less intense rainfall and slower rate of chemical weathering lead to less abrupt changes of slope profile. Frost and ice action scour and split rocks and leave characteristic residual and depositional landforms, especially in areas such as northern Europe and northern North America, which experienced Quaternary glaciations. Streams have profiles of greater equilibrium than in more arid climates.

Landforms in the arctic and subarctic zones again bear some resemblance to those of deserts. As the precipitation is lower and most falls as snow, there is little runoff and wind action on loose materials and snow surfaces is conspicuous. Physical weathering exceeds chemical and glacial and periglacial processes of erosion and deposition become dominant.

Although climatic or climatogenetic classifications have been the most widely used, the appearance of landforms also strongly reflects their lithology. Differences resulting from this cause are known as *petrovariance* and Section 3.4 has shown how they result from the way the origin of rocks is reflected in their internal structure and chemical composition.

Igneous rocks generally give landscapes whose arrangement of uplands and depressions is governed by faults and joints, the former being larger than the latter. These features are usually relatively planar, vertical, and occur in parallel groups or sets, often intersecting at right angles, or nearly so. In lava flows and sheets, joints can form columnar or palisade structures such as the Giant's Causeway or the Palisades. On a large scale, faults often determine the drainage pattern. Slopes vary in gradient, but are usually relatively homogeneous in the same type of rock under the same climate. Steepness tends to be directly proportional to both spalling size and the rate of detrital removal, so that large spalls rapidly removed leave steep slopes, whereas rocks suffering only from a surface weathering of relatively fine particles that are slowly removed have gentle slopes. Landforms on metamorphic rocks generally resemble those on igneous rocks except for the greater tendency for ridges and valleys to run parallel. This is due to the lineation of rocks that separates them into harder and softer bands. Silica is generally the most stable material so that the proportion and width of siliceous bands tend to govern the extent and relative elevation of higher ground. Decreasing hill-forming competence can be seen in the following series: gneiss—schist—phyllite—slate. Marble, which results from the metamorphosis of limestone, is massive and compact, but is sometimes broken into large blocks by joints. Although it succumbs readily to weathering in humid climates, it is relatively resistant in arid climates.

Sedimentary rocks form bedded and folded structures that differ, as has been described previously, especially according to whether they are dominantly siliceous, calcareous, or aluminosilicate. Siliceous sediments differ among themselves in resistance to erosion depending on the coarseness of their constituent particles and the degree of cementation of the matrix. In general, the coarser the materials and the stronger the chemical bonding within the matrix, the more competent the beds. A coarse conglomerate cemented within a siliceous matrix is, for instance, one of the most resistant of all rocks, whereas a weakly cemented fine sand can be almost as soft as a clay. Silt and clay have plate-shaped particles whose interlocking and cohesion permit them to form steep slopes when dry and in a nonfluid state, but their relative softness and impermeability make them especially vulnerable to gullying and fluvial erosion when wet. Another effect of the impermeability of shale and clays is to arrest downward-percolating groundwater and express it in springs along valley walls or cliffs. Limestones are very vulnerable to solution in regions of high rainfall and give rise to karstic landforms, some of which have been previously described, and often leave a surface residue of their insoluble constituents such as clay, chert, or flint. In arid climates, however, the resistance of relatively homogeneous limestones to alterations of temperature enables them to withstand the agencies of destruction better than the usually more resistant igneous and sedimentary rocks. Dolomite, being partly composed of the somewhat more soluble magnesium carbonate, can form prominent and steep cliffs and mountains. Chalk by contrast, because it is soft and highly porous, forms gentler and more undulating landscapes, although also without surface drainage.

3.6 GEOMORPHOLOGY AND SOILS

Geomorphology provides the framework on which climate, acting through vegetation, forms the soil. All sites are subject to pedogenesis—the complex of physical, chemical, and biological processes that form the soil profile with its constituent horizons. Soil is the uppermost biologically weathered part of the regolith, and can be subdivided into horizons which include the usually organically enriched topsoil and the layered horizons of the *subsoil*. On consolidated materials the depth of soils tends to increase downslope, with those in valley bottoms in humid regions often subject to waterlogging. In drier climates, the soils have less developed horizons, are low in organic activity, and in valley sites are especially subject to salinization by evaporation from shallow water tables.

Important differences in soil humus, structure, and fertility result from the base status of the parent materials. This is especially clear in the contrast between the soils derived from acid and basic igneous rocks. In general the former are lighter in color, have a high proportion of silica, are less vulnerable to weathering, and give rather infertile gravelly and sandy detritus, whereas the latter are darker, containing a higher proportion of aluminosilicate and ferromagnesian minerals, which, especially when rich in calcium, give more argillaceous and fertile soils and outwash.

Soils developed from sedimentary rocks reflect the outcrop pattern of the different strata as expressed in tabular or scarp-and-vale topography. Cliff faces tend to be lithosols. Footslopes and valleys are subject not only to mantling with unconsolidated drift but also to periodic flooding. Their soils therefore may be deeper but also contain transported material and show hydromorphic features. In humid climates they may be gleyed and peaty, in grassland climates meadowlike, and in deserts salinized.

Soils on sandstones, limestones, and shales reflect their parent materials. On sandstones they tend to be permeable, poor, and acidic, unless the rock matrix is enriched with bases. On limestones they are known as *rendzinas*. On elevated sites and slopes these are usually shallow and seasonally droughty because of the permeability of the rock, but are relatively base rich and contain siliceous and clayey residues from its dissolution. In humid climates they tend to have a well-mixed mull humus that can mobilize iron to give the soil a reddish coloration such as is seen in the *terra rossa* soils of the Mediterranean. Dry valleys and karst hollows are normally infilled with fine-textured materials, giving rise to deeper and moister soils. Shales and clays give subdued landscapes and impermeable soils. In humid climates this makes them characteristically wet, drying out later in spring and becoming saturated earlier in autumn than other soils. In dry climates they tend to form saline basins.

Unconsolidated deposits give rise to soils that reflect their geomorphic processes of origin. Glacial till and moraines are often topographically uneven and texturally varied, giving strongly contrasting moisture conditions within short distances. Boulder clays with a wide range of particle sizes contrast with fluvioglacial spreads that are better sorted, more stratified, contain a narrower range of particle sizes and are more topographically even.

Riverain alluvium generally forms low-lying tracts whose finer textures and proximity to water make the soils naturally productive, although locally subject to swamp conditions.

Sand dunes give an uneven landscape of highly permeable but base-poor and infertile soils. When mobile they are barren, but increased rainfall permits their colonization by vegetation. Their quartzitic character predisposes them to acidity, which in humid climates causes podsolization. Loess, on the other hand forms a richer, less drought-prone soil that is highly productive where topographic and moisture conditions are favorable.

The suitability of beach-derived soils for plant growth depends on their texture and freedom from salts. In general, textures become finer seawards and thus become more inherently fertile but more poorly drained in the same direction. Within this pattern, however, there is often some development of parallel tidally formed stony ridges and finer textured swales parallel to the coast.

Finally, most landforms are polycyclic, showing evidence of formation during previous periods when environmental conditions were different. In the temperate zones traces of the alternation of glacial and interglacial conditions with their effects on the vegetation and the processes of soil formation can be deduced from paleosols (ancient and buried soils), sometimes associated with identifiable pollen populations, fossils, and artifacts. Older fossil soils developed under tropical high rainfall conditions are found today in lower rainfall areas of subtropical and temperate Africa, Australia, and Latin America.

FOUR

THE VEGETAL
FACTOR

4.1 INTRODUCTION

The vegetal basis of plant–landform relationships, their interdependence, and the independent study of vegetation begins with geography or, more precisely, plant geography, which is primarily concerned with the study of large-scale distributions of plant species and communities in relation to major climatic and historical factors. However, many workers emphasized the study of vegetation as being confined primarily to small-scale distributions of plants in relation to local environmental factors, including microclimate, geology, topography, aspect, soil and soil moisture, and refer to the subject as plant ecology (see also Section 1.2).

No doubt confusion of thought can arise through confining the studies to small areas and ignoring the interplay of environmental factors and plant distributions over extensive or intermediate-sized areas. Separate studies of large and small areas have generated distinctive methods and techniques and differences of opinion between workers that on occasions may not be complementary but in conflict. In this text, no attempt will be made to present and synthesize the wide difference of opinion between workers. What seems important is to maintain a holistic approach when studying the ways in which nature is organized and to give sufficient understanding of the organization of vegetation in

46

nature so that this can be related directly to the landforms associated with the plant communities.

Previously the importance of geomorphology has been emphasized. However, except in deserts and polar regions, the land surface of the world has a green mantle of living vegetation, which becomes increasingly conspicuous as we travel from the semiarid to the wet tropics. In the tropics the profuse vegetation may make it impossible or extremely difficult to map the characteristic landforms without some knowledge of the associated vegetation types.

4.2 VEGETATION AND CLIMATE

The interaction of climate and vegetation is highly complex; and this is not surprising because climate itself, as indicated previously, is complex and often difficult to describe in exact terms. The major components of climate influencing the development and permanency of plant communities are incident solar radiation, ambient temperature, precipitation (and mist), atmospheric humidity, and wind. To these must be added other environmental factors affecting the distribution of plant species and the physical structure of the vegetal types. These include landforms, soil conditions, water availability, aspect (exposure), and the influence of precipitation and temperature on evapotranspiration. Precipitation and the availability of water for plant growth are often correlated with altitude and aspect. A new dimension is provided when seasonal variations in climate are taken into consideration. In agriculture this cumulative factor is often expressed as the number of crop-growing days needed for the economic production of specific crops in a geographic region (Higgins and Kassam, 1981).

Various workers have helped to demonstrate the correlation of vegetation types and plant species with changes in microclimate and regional or macroclimate. Schimper, nearly 100 years ago (1898), was the first major worker to recognize the close relationship usually existing between climate and vegetation and to coordinate major differences in climate with vegetal types. He proceeded to classify vegetation types (e.g., forest, woodland, grassland) according to whether the climate was wet, seasonally wet, or dry and further classified the vegetation on a broad physiognomic basis. Thus the Australian forests were divided into sclerophyllous (with hard leaves and usually xeromorphic) and rainforest (having mostly mesomorphic or hydrophilous leaves). The subtropical forests dominated by *Araucaria* were regarded as impoverished tropical rainforest.

Later Clements (1916) emphasized the importance of climate in influencing successional vegetal changes and concluded that climate controls the vegetation to such a degree that vegetation within a climatic region evolves towards a climatic climax unless constrained by other factors including fire and human influence. However, in recent years the importance of factors other than climate, particularly endemic fire, has also been increasingly recognized as providing the development of stable vegetation (e.g., Australian eucalypt forests, savanna woodland in Africa). In the Mediterranean zone, including parts of California, Australia, South Africa, and Chile as well as the type area, the climatic evergreen forest, once it has been disturbed by human activities, quickly degrades from forest to maquis, to garigue and to steppe, and is very slow to recover.

Most frequently single climatic factors have been used to explain major differences in adjoining vegetal types. For example, Köppen (1931) observed that in the northern USSR the timberline coincided with the 10° July isotherm. Similar correlations between temperature, vegetation, and altitude can often be observed, particularly in the tropics. Troll (1943) observed sharp changes in the vegetation types of the Peruvian Andes where the 0°C isotherm occurred. In the Sahel (west Africa) there is a marked correlation between natural vegetation type or agricultural crop and the increasing rainfall in a southerly direction.

The relationship between vegetation and rainfall and temperature is probably best seen in mountainous regions of the tropics as one ascends from the tropical lowlands to the tropical alpine climate at high elevations and as one moves from the slope facing the rain-bearing winds to slopes in the rain shadow. The tree line where forest gives way to grassland or shrubland can be observed for similar aspects at approximately the same altitude on isolated volcanic mountains in east Africa. For the same elevation on Mount Meru in Tanzania, montane tropical forest occurs on the southerly and easterly slopes and coniferous forest (*Juniperus procera*) occurs on the drier northerly and westerly slopes. In some tropical countries the preparation of small-scale mean annual rainfall maps has relied considerably on the boundaries of vegetation types to extend the boundaries provided by localized rain gauge readings (e.g., Tanzania and Zambia).

In the mountainous areas of temperate regions and the lower rainfall mountainous regions in the tropics, the distribution of vegetation types can often be observed to be influenced by aspect as associated with exposure, wind, and insolation. In Colorado shrubs and grasses occur on the low southerly slopes of the Front range, whereas forest dominates the northerly slopes. On the slopes of the Grand Canyon, major transitional vegetation types can be observed within a few hundred meters, mainly because of the aspect–insolation effect. At higher latitudes the change with aspect in vegetation types is often striking between northerly and southerly aspects.

In Australia, Petrie, Jarrett, and Patten, as early as 1929 drew attention to exposure influencing the distribution of wet sclerophyllous forest, which can attain a height of about 100 m. The presence of dry sclerophyllous forest on western slopes of the Dandenong Range (Victoria) is associated with shallow soils and the dry northerly winds; and on sheltered slopes the soil is deep and occupied by wet sclerophyllous forest. When the annual rainfall is below about 1000 mm, the sclerophyllous forest is replaced by a woodland containing widely spaced, large-crowned, short-boled trees and an understory of shrubs may be well developed although short in height. When the annual rainfall is below about 625 mm, the woodland contains taller but narrow-crowned very widely spaced trees and a discontinuous shrub layer.

4.3 VEGETAL TYPES OF MAJOR CLIMATIC ZONES

As indicated in the previous section, there is usually an observable correlation between broad vegetation types and their associated macroclimate, and therefore it is not surprising that an initial division of the world's vegetation into types or classes usually takes into account the climatic zones in which the vegetation occurs, and further subdivision is based on plant physiognomy, floristics, or habitat characteristics. Alternatively, vegetation types extending over large areas and possibly entire geographical regions may be characterized directly in terms of their gross physiognomy. Du Rietz (1936), in applying a physiognomic approach, termed these extensive vegetal types panformations (e.g., boreal forest of eastern Canada).

In this section, vegetal description will be introduced by considering briefly the vegetation occurring in somewhat arbitrarily defined climatic zones of the world. Major climatic zones of the world and their associated vegetation types are shown in Figure 4.1.

4.3.1 Arid Zones

The rainfall is usually less than 200 mm a year and evaporation exceeds precipitation, thus causing a water deficiency. Therefore the vegetation is sparse or absent and all local streams are intermittent. The soils are light in color, low in organic matter, and usually sandy. Because of high evaporation, concentrations of salts and alkali are common in desert basins. As the rainfall increases the areas become less arid and support vegetation where the water table is near the surface or the land is irrigated. Tree roots have been found at depths of about 80 m.

The main characteristic of the natural vegetation is the lack of continuous plant cover and usually a small range of species. Clumps of grass, bushes, and cacti are scattered. Within these areas the gradations of soil moisture are accurately indicated by degree of sparseness of the vegetation. The vegetation is denser and different species appear where there is more water. Channels of intermittent streams are emphasized by vegetative growth. Water-bearing joints

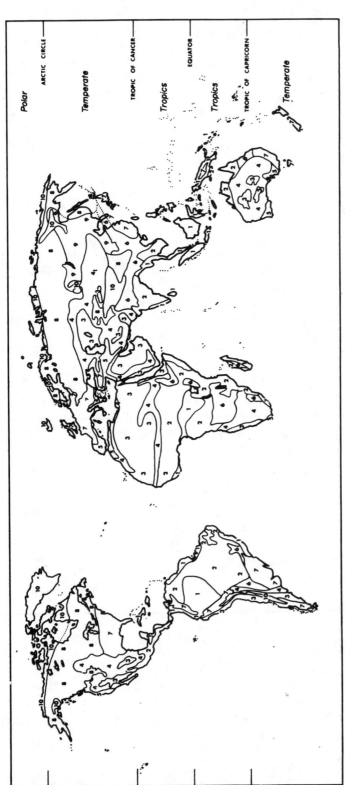

Figure 4.1. Climatic zones of the world (after Köppen, 1923), linking climate and plant panformations: 1, humid tropical zone (rain in all seasons—tropical rainforest); 2, semihumid tropical zone (distinct dry season—high forest to savanna); 3, arid climate (no annual rainfall); 4, semiarid zone (seasonal low rainfall—Sahel, steppe, veld); 5, warm temperate zone with hot dry summers (Mediterranean type vegetation); 6, warm temperate zone with dry winters; 7, temperate zone with rain in all seasons (includes deciduous forest/woodland); 8, cool temperate zone with precipitation in all seasons (cold winters with snow—coniferous forest/woodland); 9, cool temperate zone with dry cold winters (coniferous forest/woodland); 10, polar climate (cool/cold in all seasons—tundra).

or faults are often noticeable by a linear vegetative trend. In sand dune areas, the interdune tracts often support scattered plants, indicating that vegetation establishes itself quite rapidly where water is available. Interbedded sedimentary rock patterns are emphasized where one of the rocks is an aquifer (sandstone) and the alternating layers of rock are shale. Bands of vegetation appear on the lower contact of the sandstone with the shale, often looking like contour lines. Small bushes will frequently form an irregular bordering band of denser growth outlining the toes of alluvial fans where the water is nearer the surface. *Phreatophytes*, so-termed because they draw water from the water table, may be common (e.g., mesquite and salt cedar in the United States).

4.3.2 Semiarid to Subhumid Zones

Rainfall usually exceeds 200 mm and may extend, if seasonal, to about 1000 mm or occasionally 1200 mm. These zones form a broad belt between the arid and humid regions. They are distinguished from arid zones by their ability to support continuous vegetation. They differ from the humid regions in that they do not receive enough water to support a continuous cover of high forest, but tall trees grow along stream banks and on flood plains even where the climate is semiarid. As the rainfall increases (e.g., Sahel in Africa) and in the higher rainfall areas of the tropics, savanna shrubland and woodlands may predominate and gallery (riverain) forest may occur along major water courses. Under lower temperatures with a winter rainfall maximum such as around the Mediterranean, similar rainfall is more beneficial than in the Sahel where temperatures are always high and the rain maximum is in summer. In the United States the gradient is from sagebrush through grassland to deciduous woodland.

The major differences in moisture are reflected in the species of grass and shrubs and their patterns of growth, both locally and regionally. The drier parts of this zone have the short grasses, which tend to grow in clumps. The wetter part of the region is continuously covered with tall grass or woodland. Each vegetation type usually has its own associated soils. Alluvial and eluvial soil boundaries are often clearly defined by the vegetation.

In the lower rainfall areas, salts and alkali in the soil are a problem, as in the semiarid regions and the desert, causing reactions that occasion local changes in environmental conditions. These in turn are reflected in changes in the native vegetation (and agricultural crops). The similar ecological conditions occurring in different parts of the world often result in vegetation types with similar physiognomy. Examples of such *homologues* include maquis (Corsica) and chapparal (California); prairies (central United States) and steppes (USSR); and sagebush (United States), wormwood formation (Caspian, USSR), and saltbush (Australia).

4.3.3   Temperate Humid Zones

Precipitation exceeds evaporation, and there is sufficient moisture available for the support of extensive forests. The zone is subdivisible on the basis of mean annual temperature into cool and warm temperate. Often the forests have been completely removed to make way for farms and homes. In such areas, in contrast to the drier regions, trees are commonly found in woodlots, along roads, and so on. The character of the soils in these regions varies greatly, depending upon the associated landforms and the variations of parent material within a landform unit. In general the soils are deep and have considerable quantities of organic matter especially in the cooler areas. Associated with swamps and marshes are the peaty organic soils. In the cool temperate zone the landform, soils, and indigeneous vegetation may have been influenced by periods of glaciation.

Of the wide variety of vegetation found in humid regions, the most prominent are the woodlands and forests. For example, the names of the major North American forest types (panformations) are (a) boreal, (b) western, (c) eastern (with its subclassification of northern mixed, central deciduous, and southeastern mixed), and (d) subtropical. Within each of these broad types, we can recognize smaller types of plant formations. The western forest type is divided into the Pacific coast forest, the Sierra Nevada forest, and the Rocky Mountain forest, and each of these forests has its own subdivisions.

The commonest way of classifying the forest is by the kinds of trees that make up the forest canopy. Needleleaf trees provide the coniferous forests; broadleaf trees form winter bare or deciduous forests; coniferous and deciduous trees growing together form mixed forests. Mountain forests may be classified according to the horizontal zones they occupy. The forest on the lower slopes of a mountain is often called foothill or *submontane forest*; the forest on the middle slopes is called *montane forest*; and the forest on the upper slopes, *subalpine forest*. The upper margin of the subalpine forest where the trees become dwarfed or are absent is the timberline. Latitudinally the forest gives way to the treeless tundra, bare rock, snow, and ice. The elevations of these zones vary with the latitude, wind velocity, and aspect.

Frequently the names of the tree species that are predominant in a forest or woodland are used to identify the forest type or association. Thus in the eastern United States there are the oak–hickory forest type, the beech–maple forest type, the southern pine forest type, the hemlock-white-pine-northern-hardwood forest type, and others.

As a general rule the height of the vegetation is moisture-dependent. From the well-drained soils of table land to steep hillslopes and of a hillside to the open water of a pond, the average height of the vegetation decreases. From the tall trees of a *climax* forest of a gentle hillslope, one may travel through successive zones of shrubs, sedges, and grasses, swamp plants floating rooted plants, underwater rooted plants, and finally, floating plants. All of these *seral stages* may be seen in the area where ponds are drying up and/or filling in with vegetation. As the pond continues to fill in by natural succession, deciduous trees may overgrow the evergreen forest zone, the evergreens will overgrow the shrub zone, and so on. These stages do not always occur, however. If the swamp conditions are relatively permanent, extensive swamp forests will be the climax community.

4.3.4   Alpine Zone

Alpine grassland (alpine meadows) occurs at high elevations in the humid temperate zone and sometimes at very high elevations in the tropics. That average temperatures and temperature ranges decline with increasing elevation is well known, but the altitudinal effect on precipitation is not nearly so well documented. The effect of the interaction of these two factors on vegetation can often only be surmized. In temperate regions the timberline that defines the lower limit of the alpine grassland and the upper limit of the subalpine forest may vary in altitude by more than 1000 m.

The climate of alpine grassland is characterized by a short cool growing season, ample moisture, and long periods of snow cover in winter, often associated with strong winds and extended periods of cloud or mist. The transition from forests to alpine grassland may be quite abrupt. The alpine meadows above the timberline are often highly productive as compared with the forest immediately below the timberline. Consequently, in Europe, for example, farmers have extended their summer grazing lands (alpine meadows) downwards by clearing the forest.

4.3.5   Humid Tropical Zone

This zone generally has the necessary temperature and rainfall for year-round growth. The rainfall usually exceeds 2000 mm and may exceed 5000 mm (e.g., Sierra Leone). Exceptions occur in the high mountainous areas where vegetation types occur similar to those in humid temperate zones. Richards (1952) drew attention to the logic of beginning a system of vegetal classification with tropical rainforest because therein the environmental factors are least limiting.

In general the vegetative cover is luxuriant and continuous, but major variations occur due to the local differences in landforms, temperature, precipitation, soil permeability, soil fertility, human activity, and so on. The influence of parent rock material must also be taken into consideration. For instance, basic igneous formations are much more favorable for vegetal development than serpentines. Age of the rock type is another consideration. Tropical conditions are ideal for rapid soil formation. It is usual for granites to be weathered to depths of 70 m or more. These soils are often lateritic, that is, enriched in iron and aluminum and leached of the more soluble plant nutrients (e.g., silica). Older soils may become so leached that they cannot support tall or closely spaced trees. Also, lateritic crusts and gravels support little vegetation.

Rainforests are prominent in tropical South America, western Africa, southeast Asia, and the tropical Pacific. They occur over extensive areas and also in less favorable locations, along high banks of water courses, and on the slopes of mountains, often below the moss forests. The undergrowth of disturbed rainforests consists of tangled shrubs, creepers, vines, smaller trees, bamboo, and other plants up to 10 m high. Mature rainforests with their dense canopy and abundance of tree species have little undergrowth. The canopy is several layered with frequent large emergent tree crowns. Under conditions favoring less luxuriant growth the rainforest is replaced by lowland high forest or on higher ground by submontane or montane forests (e.g., moss forests).

After clearance for agriculture, which because of population pressure is increasing markedly, the soil fertility often deteriorates rapidly. The forest stand height at maturity probably will not exceed 45 m. Leaves develop drip tips, plank buttresses are common, and the crowns with many epiphytes are often interlaced with lianas.

Moss forests are widely distributed in the tropics where mountains are covered by an almost continuous cloud belt and rainfall is high. This type of forest is most common on steep mountain slopes between altitudes of 1000 and 3500 m and slightly lower on windward sides. Savanna and occasionally desert conditions are often found on plains to the leeward of mountain ranges. The soil and deep humus are completely saturated by the long periods of rain and fog. The soils are generally clays or sandy clays, depending upon the parent material. The stand consists of a dense growth of small trees, often less than 10 m high, with branches close to the ground dripping with moisture. The tree trunks, tangled branches, and ground surface are covered with mosses, ferns, and lichens.

These moss forests are somewhat similar to rainforests but are not multilayered and neither so tall nor so rich in species. Crowns of individual trees present a slightly distinct round or conical appearance and are much smaller than those of rainforest. Mature moss forests have few emergent crowns.

Swamp forests are found in the tropics on low swampy land, adjacent to streams, which are subject to flooding during the rainy season and often comparatively dry at other times of the year. Swamp forests are frequently found in association with scattered palms and adjacent to rainforest occupying higher ground. They consist of a dense growth of trees and undergrowth that are similar to regrowth rainforest. A distinctive feature of the swamp forest is the large buttressed trunks common to many tree species. The tree crowns are often flat and/or engulfed by vines. The cover is not as interlaced as in rainforests. Water is usually present, but sago or nipa palms are common in some areas. Swamp forests cannot tolerate salt or brackish water.

Wild cane in the tropical Pacific region, up to 4 m high, occurs in swampy low-lying coastal plains, lake plains, or river flood plains. It is often found along stream banks or overflow areas, and is usually confined to small sites that may be covered with several feet of fresh water. Typical conditions find it adhering to meandering stream courses. Tall grasses occupy slightly higher portions of the flood plain, whereas short grasses are found on interstream ridges.

Palms mostly occur in inland swampy areas beyond the extent of tidal water, particularly in deltas and flood plains. Palm stands are typically found adjacent to, and lower than, swamp forests (and occasionally the rainforest) bordered on the seaward side by nipa palms. Sago palms are usually more uneven in height, 10 m or more, whereas nipa palms rarely exceed 5 m in height. Pure stands of sago palms are less common than nipa palms, for sago usually grows in association with other palms and hardwoods. Nipa palms are distributed throughout the tropics on sea level sites that are usually submerged at high tide. They often fringe tidal streams, but are less tolerant to salt water than mangrove and in general are limited to brackish waters. The nipa areas often adjoin swamp forest, rainforest, or sago areas. The stand height is usually 3–5 m and the palm forms such dense stands that few, if any, other plant species grow in association with them.

Mangrove swamps occur along tropical, subtropical, and occasionally temperate coast lines between the limits of tide and along stream banks as far inland as the limits of brackish water. The trees in the tropics usually grow to a height of about 15 m, although occasionally they reach heights up to 20 m (e.g., *Rhizophora* in Bangladesh and Tanzania) and have thick, dark green leaves and exposed aerial roots. The roots may extend from the trunk at a height of 3 m or more above the ground and develop at right angles to the trunk. Because the trees grow close together, the roots and trunks form a dense tangle. The small mangrove species (*Avicennia*) are often found at the outer fringe of the mangrove swamp. Upstream edges may be bounded by nipa palms, and some larger trees, including the mangrove *Bruguieria*, are often found on the inner edge of the swamp where it merges into the normal swamp forest or rainforest. At low tide the ground is often thick mud from a few inches to several feet deep. Numerous small, but often deep, water courses meander within the general swamp. In the subtropics mangrove swamp usually gives way to tidal marsh although dwarf mangrove occurs as far south as 38° in southeastern Australia.

Grassland in the tropics indicates either a relatively dry environment, frequent burning, or edaphic conditions unfavorable to tree growth. Soils in permanent grassland areas may vary from laterite or lateritic soils to young soils. The grass itself is often shaded or interspersed by trees in savanna areas. The short grass and savanna areas are found from sea level to medium altitudes. A typical association shows short grasses growing on rolling to steep hills with savanna on a rolling to flat plain at the base of the hills or mountains—often a pediment surface. Tall grasses are an exception to the dry soils rule; they grow on plains, plateaus, and occasionally rolling hills at lower altitudes and are often associated with seasonal burning; they also indicate wetter soils and seasonal swamp areas.

4.3.6   Arctic and Subarctic Zones (Polar)

In general, this is latitudinally a zone of low rainfall (125–400 mm annually), but the ground is often wet because evaporation is low. The tundra vegetation in the Arctic is due to the effect of climate and the presence of permafrost and is independent of soil texture. Only the top 1 or 2 m of permafrost in the soil areas usually thaws in the summer, although this is deeper in basin sites where superincumbent water may lead to deeper melting and the development of thermokarst. Vegetation consists of grasses, mosses, and widely scattered stunted willow. Mosses and lichens grown on exposed rocks.

The thick moss cover combined with the generally low relief combine to retard runoff, making the region very marshlike. Below the 32°F average line (the sub-Arctic), permafrost gradually becomes discontinuous, and scattered, stunted small trees begin to appear in the valleys and protected depressions. Farther south, as the soil cover develops and the climate is less severe, stands of the boreal forest develop.

In the sub-Arctic the type of vegetation is often useful in delineating areas where permafrost is likely to be present; shallow-rooted species indicating permafrost occuring at shallow depths and deep-rooted species indicating permafrost at greater depths or absent. For example, stabilized sand dunes of interior Alaska support stands of fairly dense black spruce in contrast to the brush and grass of the interdune areas. This relation is found where the higher, well-drained sand of the dunes contrasts with the frozen silty deposits of the interdune areas. However, black spruce may also indicate permafrost at shallow depths. Thus, although vegetation alone should not be used as an indicator of permafrost conditions, willows often indicate a low permafrost table and the likelihood of ground water. Thick, peaty surface layers act as insulators and aid in development and preservation of permafrost. Where the natural insulation is the thickest, the permafrost table is ordinarily nearest the ground surface. Some types of tundra and muskeg are good indicators of permafrost close to the surface.

4.4   PHYSIOGNOMIC BASIS TO VEGETAL CLASSIFICATION

Following Clements (1916), Holdridge (1947, 1966) has demonstrated in Latin America the importance of climate in classifying the vegetation covering extensive areas into meaningful major vegetal units that can be described in terms of their physiognomy. He has successfully combined rainfall, temperature, and potential evapotranspiration in order to divide the Andean vegetation into plant formations. Holdridge's quantitative scheme is worldwide in its approach, provides for 38 basic vegetation types, and clearly links climate and vegetal structure. The total range is from hot deserts and polar ice sheets to tropical rainforest. The basic method is summarized diagrammatically (Figure 4.2) and consists of three parallel lines inclined to each other at 60° in the form of an equilateral triangle. One set of parallel lines represents annual precipitation in millimetres (125, 250, 500, 1000, 2000, 4000, 8000) and the second set shows the mean annual temperature in degrees centigrade (0, 3, 6, 12, 24). The third set indicates evapotranspiration values (0.25, 0.50, 1.0, 2, 4, 8, 16). As shown in the figure, these values are also linked to altitudinal and latitudinal zones. Within these types the plant formations are described in terms of their physiognomy and can be subdivided into plant subformations, whose distribution is influenced by the landforms and soils.

4.4.1   Terminology

Difficulties in the use of terminology also occur when considering the physiognomy of vegetation. The term *physiognomy* will be used to embrace the external shape or form of the stand (i.e., external physiognomy), which is so readily observed on stereoscopic pairs of aerial photographs, the internal structure of the stand, and the shape or life-forms of the dominant species of the stand. Küchler (1967) used as his physiognomic criteria life-forms, structure, and seasonal periodicity.

The identifiable changes in the physiognomy of vegetation result from the heterogeneity between plant communities and homogeneity within the plant communities. Opinions differ greatly on the importance by area of *ecotones* or transition stands, which occur between discrete communities and, depending on this, workers favor either the classification or ordination of plant communities. For example, a line drawn on an aerial photograph to represent a change in forest types could imply that the plant assemblage to which it relates had discrete boundaries and that continuum theory is rejected; or only that it is of practical convenience in forest typing. Some authors (Clements, 1916; Tansley, 1920) have implied that communities have more or less sharply defined

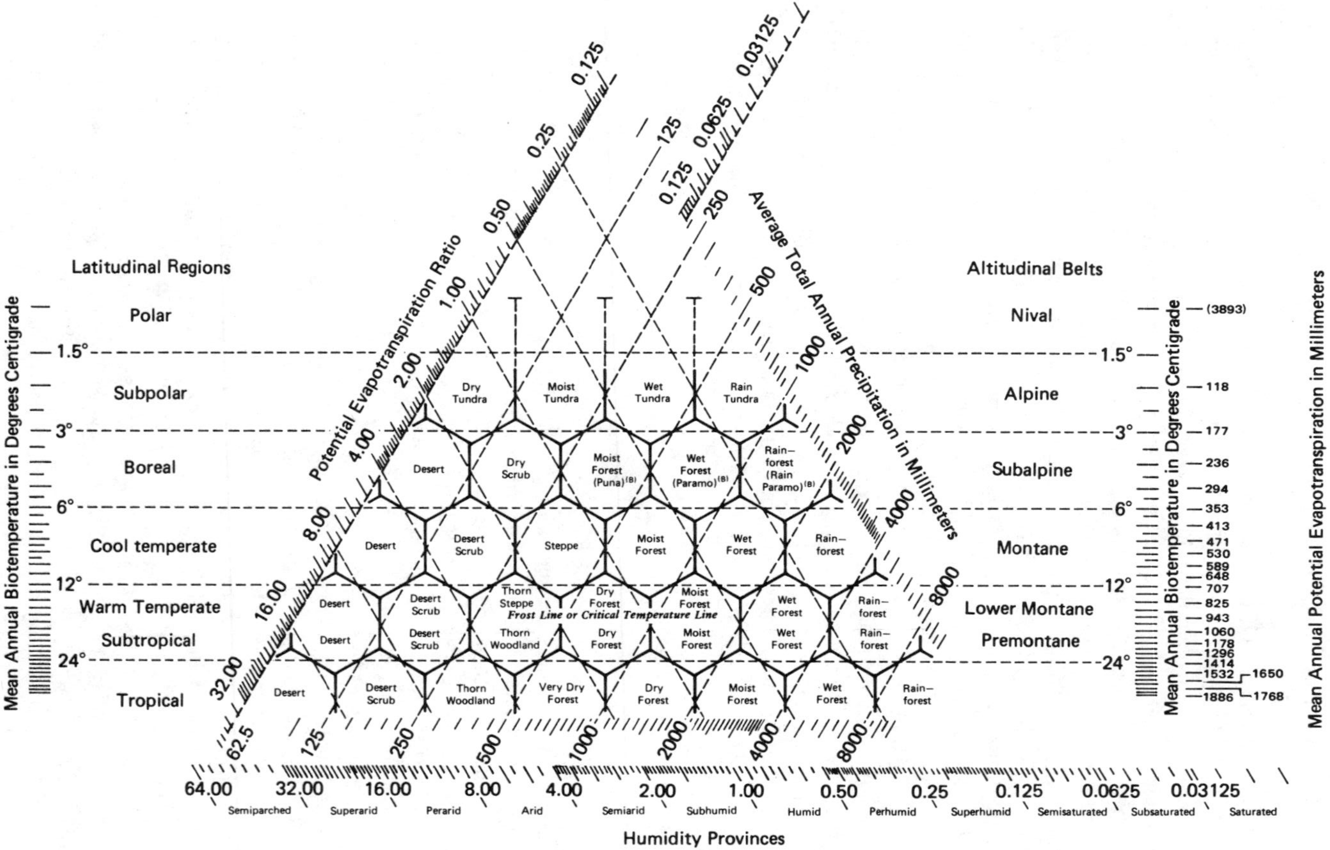

Figure 4.2. The linking of climate with the vegetal structure of major plant formations by using average total annual precipitation, mean annual bio-temperature, and potential evapotranspiration ratio (Holdridge, 1966; Holdridge and Toshi, 1972).

58

boundaries, whereas others (Curtis, 1959) have demonstrated vegetation to be in a continuum.

The term *classification* has come to imply, through publications, that plant communities have discrete boundaries and their composition, whether physiognomic or floristic, is discontinuous. *Ordination* on the other hand implies the overall importance of continuous variation in the composition of the plant communities, although not denying the occurrence of discontinuity. In practice the results provided by the two approaches may not differ so greatly. The reason for this is that the purpose of the study is usually the same or similar, although it is achieved by different methods and possibly by using different characteristics of the community. In ordination it is assumed that the vegetation exists in a continuum and an attempt is made to place each stand in correct perspective to other stands usually in relation to graphical axes. In classification, boundaries are drawn where they are observed to exist or in such a way as to minimize the influence of intermediate stands (ecotones).

Many workers recognize the plant *formation* as their highest ranking plant community. This is a structural unit to which are referred all climax communities exhibiting the same structural form. In a *climax community* the species perpetuate themselves through natural reproduction; it is thus the final step in plant succession (Mueller-Dumbois and Ellenberg, 1974). A *subformation* or order (Braun-Blanquet, 1951) is a subordinate division of the plant formation on the basis of structure and not of the ground area occupied by the structural unit.

Formations and subformations have not only a characteristic internal structure that can be portrayed in profile diagrams (e.g., Dansereau, 1951; Mills et al., 1963), but also have their own characteristic three-dimensional surface profiles in the stereoscopic plan view provided mainly by the tallest stratum recorded on the aerial photographs (Howard, 1959). The latter was termed *photophysiognomy* by Howard (1970c). The profile represents in pictorial form the organizations in space of the individual plants composing the vegetal type. Tansley (1953) regarded the tallest stratum as giving "the community its characteristic physiognomy or appearance and that this largely controls its structure."

Both the stereograms as provided by the overlap areas of a stereoscopic pair of aerial photographs and the plant profile diagram prepared from field measurements along a line or strip transect (on a fixed compass bearing), are essential descriptive components of the physiognomy of the vegetation. Each portrays the vegetation that exists and does not presuppose any special relationships with the edaphic and climatic conditions. The profile diagram (Figure 4.3) provides a two-dimensional representation of the three-dimensional view of the vegetation as observed and measured on the ground. The horizontal dimension represents the area covered by each life-form, whereas the vertical dimension represents the height or height class of the life-forms.

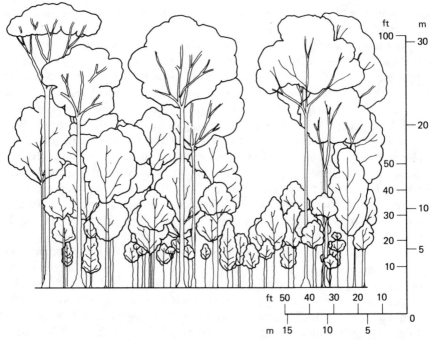

Figure 4.3. Profile diagram of tropical forest. The horizontal dimension represents the area covered by the life forms in a strip approximately 8 m × 40 m. The vertical dimension represents all trees present in the strip with a height exceeding 4 m. Note the presence of tree buttresses and the characteristic crown shape of the trees in the main canopy.

4.4.2   Physiognomic Characteristics

Plant communities have many characteristic features, including their floristic composition that will be considered later. Floristic features increase in importance as the vegetal classification studies and mapping procedures become more intensive and restricted to smaller and smaller areas. Even when attention is confined to the physical characteristics of vegetation, it is not possible to use all the physiognomic characteristics and it is necessary therefore to be selective.

Classification methods embody floristic characteristics, habitat characteristics, and the physiognomy of the vegetation. The divergences reflect the object of the work, the interests, training, and experience of the worker(s) and the environment. For example, Braun-Blanquet (1951) considered that a physiognomic classification was not precise enough for ecological studies in continental Europe; whereas Poore (1963) thought it embraced faithfully the ecological factors of the habitat, and Beard (1944) concluded that the physiognomic characteristics of formations were adequate for their classification.

The physiognomic classification of the vegetation has been achieved either by synthesis of the smaller plant floristic units (subassociations and associations) into larger floristic units (alliances) and then into the physiognomic units of subformations and formations or the formation and subformations have been identified directly as physiognomic units (Coaldrake, 1961; Wood and Williams, 1960). The number of structural expressions of vegetation can be considered as relatively few compared with floristic groupings, which may be viewed for extensive areas as virtually unlimited.

The three main physical characteristics of the vegetation, which are used in physiognomic classification and which can be accurately measured and mathematically defined, are mean height of vegetation, the ground cover of the vegetation, and the life-form of the vegetation. Typical forest life-forms, for example, are conifers (gymnosperms), deciduous hardwoods, sclerophyllous forest (hard-leaved), and rainforest (hygrophilous) (Schimper, 1903).

Many workers have preferred to use plant cover as the first criterion for separating vegetation into types (*UNESCO*, 1973), although there is growing recognition of height as being of primary importance (Howard, 1970c,d), because it is the more "robust" and not so strongly influenced by human activities. *Height*, or more precisely *stand height*, is usually determined in the field for forest vegetation by either the measurement or visual assessment of a reasonable number of dominant trees, codominants, a mixture of both, or occasionally for each sample plot all trees including subdominants. The same approach can be used for shrubs where trees are absent.

Vegetation or *plant cover* is defined as the proportion of the ground covered by the vertical projection on to it of the overall vegetation canopy. The numerical result is expressed as a percentage (e.g., 80%), often termed *crown cover* or *crown closure* for woody vegetation, or as a decimal coefficient (e.g., 0.8) that is termed *crown density*. If the vertical projection represents the overall crown, then this is more precisely *gross crown cover*; but if it represents the vertical projection of the individual aboveground parts of the plants (i.e., leaves and branches), then the term *net crown cover* should be used. If the total horizontal leaf area is summed and divided by the ground area, the resultant is termed *leaf–area index*. Leaf–area indices of 4 to 8 are common for the vegetation of the humid temperate zone. There are no generally accepted plant cover classes, but frequently vegetation may be classified based on cover as follows: over 80% (closed), 40–80% (dense), 10–40% (open), 2–10% (sparse), and under 2% (absent or rare). Plant cover should not be confused with *plant biomass* (phytomass) per unit ground area, which refers to the total dry weight per unit area.

Raunkiaer (1934) was among the first workers to give considerable attention to height as a basis to separating vegetal life-forms, although height is traditionally used by forest managers. He recognized megaphanerophytes (over 30 m in height), mesophanerophytes (8–30 m in height), microphanerophytes (2–8 m), nanophanerophytes (0.25–2m) and chamaephytes (less than 0.25 m). Plant formations are usually readily identified according to life-forms and stand height as provided by Raunkiaer (1934). Howard (1970c) using Raunkiaer's life-forms developed a two-way classification based on the photographic interpretation of stereoscopic pairs of aerial photographs to describe the plant formations and plant subformations of eastern Australia and eastern Africa (Table 4.1). Canopy closure of the woody vegetation was used as the second parameter. Küchler (1967) has also used height as the first parameter and plant cover as the second for the stratification of the natural vegetation into types.

By choosing appropriate height classes and canopy-density–vegetal-cover classes, important plant formations can be identified and described. For example, mature vegetation may be divided into several height classes, although there are always exceptions and appropriate descriptive terms assigned to describe each of these classes. We may describe vegetation over 30 m as forest, 10–30 m as woodland, 2–10 m as shrubland, 1–2 m as tall grassland, and under 1 m as low shrubland or herbland–short-grassland.

Beadle and Costin (1952) in Australia, influenced by Raunkiaer, defined a *forest* as a closed community, dominated by mesophanerophytes (8–30 m in height) and megaphanerophytes (> 30 m in height) that have flattened crowns, form normally an interlacing canopy, and have a bole–length ratio usually > 1. Woodland was defined as a community dominated by crown length mesophanerophytes with rounded crowns, an open or loosely interlacing canopy, and a bole–length ratio usually < 1; scrub as dominated by single-stemmed crown length microphanerophytes (2–8 m in height); *heath* as a closed community dominated by nanophanerophytes (shrubs 0.25–2.0 m in height) and chamaephytes (< 0.25 m in height). Finally, *savanna* was recognized as dominated by perennial xeromorphic grasses and widely scattered woody phanerophytes and *grassland* as dominated by perennial grasses.

Plant formations based on climate and gross differences in the physiognomy of the dominant layer can be further divided into plant subformations either in the field on the basis of their stand structure recorded as profiles along line transects or by using their aerial photographic parameters to describe the subdominant layers and to provide additional information on differences in the dominant layer (Howard, 1970c, Howard and Schade, 1982). Thus the vegetation type as a formation may be described as temperate woodland; but on the characteristics of the subdominant layer or ground layer of vegetation and major differences in the dominant layer can be divided into several subformations.

Table 4.1. An Example of Two-Way Physiognomic Classification of Vegetation Using Primarily Dominant Height and Gross (Ground) Cover of Woody Vegetation.

Life-Form	Megaphanerophytes	Mesophanerophytes	Microphanerophytes	Nano-phanerophytes	Chamaephytes
Height (meters)[c] Physiognomy	>30 Forest	8-30 Woodland	2-8 Shrubland	0.25-2.0 Herbland, heathland Heath	<0.25 Herbland
Gross cover (dominant woody strata) ~100%	Rain forest (e.g., tropical, temperate)[a]				
Gross cover dense, >50%	Closed forest (e.g., tropical, wet montane[b] Wet sclerophyllous forest[a]	Closed woodland Layered woodland[a] Tall woodland[b]	Thicket (e.g., coastal swamp[a])		
Gross cover 10-50%	Open forest[b] Dry sclerophyllous forest[a]	Shrub woodland, Low woodland Alpine woodland[b] Savanna woodland Wooded heath[a] Wooded grassland[a] Swamp woodland[a]	Thicket (e.g., mallee[a] itigi[b] Shrub grassland	Steppe (e.g., saltbush)[a]	
Gross cover open <10%		Scattered tree savanna		Tall grassland[b]	Swamp; hummock grass cover[a] Tussock grass cover Salt marsh

[a]Based on observations of formations/subformations in eastern Australia.
[b]Based on observations in east Africa (Tanzaniak).
[c]Sensu Raunkiaer. 1934.

These might include closed woodland (dense), savanna woodland (open), layered woodland (understory of small trees), shrub woodland (understory of shrubs), heath woodland (with an understory of heaths), and grassy woodland (with a graminaceous understory) (see Tables 4.1 and 8.2 and Figure 8.8). Furthermore, the habitat (topography) may be the major limiting factor (e.g., swamp) that can be combined with the dominant species (e.g., *Sphagnum* swamp, *Molinia* swamp).

Obviously, many natural communities are influenced or replaced by humans to provide induced vegetation types. Anthropogenic-induced vegetation can be readily described in terms of its physiognomy. It is therefore often important when starting to classify the landscape into land-use categories to begin with the two broad categories of *natural vegetation* and *induced vegetation*. Induced vegetation types based on its physiognomy include arable, pasture, rough grazing, scrub, orchards, plantations (such as forest, coffee, tea, rubber, palm oil) and possibly forest regrowth (e.g., bush fallow in Africa, taungya in southeast Asia, coppice and coppice with standards in Europe).

The dynamic qualities of the vegetation can be further expanded by recognizing its primary or secondary stage of succession and identifying its seral stage (e.g., hydrosere) toward the climatic climax. This obviously makes a study of the vegetation much more complex and time consuming, but may be essential. For further information, a suitable text should be consulted (Küchler, 1967; Mueller-Dumbois and Ellenberg, 1974).

4.5   FLORISTIC CLASSIFICATION

The physiognomic classification of vegetation can lead into or be combined with floristic attributes of the vegetation. For example, in forest surveys, it may be expedient to divide the natural forest into major physiognomic units (formations, subformations) and then further describe their floristic subdivisions in terms of their dominant or economic tree species. Alternatively, as strongly favored by many plant ecologists, the vegetation is divided into plant communities on the basis of the presence or absence or dominance of the plant species (without consideration of the physiognomic classification) that are identified according to the universally accepted binomial classification of Carl von Linné. An intermediate approach is pursued by the University of Toulouse in their *Carte International du Tapis Vegetal*, which portray the major vegetation types according to their physiognomy and dominant species, taking into consideration the bioclimate, soils, the Yamgaombi physiognomic classification of plant communities, and other available ecofloristic data (F. Blasco, personal communication, 1984).

Plant ecologists have differed greatly in defining the term *plant community*. On the one hand, as a general term the plant community has been viewed simply as a group (or list) of species growing together. The group may be large or small or of indeterminate size. Frequently the plant community is viewed as the smallest recognizable floristic unit of vegetation. This unit may have recognizable common tolerances but with the individual species having little or no effect on each other; or the individuals within the plant community can be considered to function for their mutual benefit as an organism or quasi-organism (Clements, 1916; Tansley, 1935; Curtis, 1959).

Some plant ecologists have used the term *plant association* to describe the basic floristic unit of vegetal classification. These include Tansley (1949) in the United Kingdom and Braun-Blanquet (1932) in continental Europe. Beadle and Costin (1952), for example, defined an *association* as a climax community "of which the dominant stratum has a qualitatively uniform floristic composition and which exhibits a uniform structure as a whole." The floristic composition used in identifying an association is thus confined to the dominant or tallest stratum; and this may be compared with Wood and Williams (1960) who recognized associations by the dominant species of each stratum.

A floristic subdivision based on other than the dominant stratum has been termed a *subassociation* (*sociation* by Du Rietz, 1936). A subassociation is determined by a variation in the most important subordinate stratum of the association without significant qualitative change in the dominant stratum. It is often necessary when examining vegetation to group together two or more closely allied plant associations as a *plant alliance*, because changes in their recognizable floristic composition are too subtle to permit the separation of distinct associations. The grouping of plant associations or plant alliances into subformations does not normally produce difficulties, because structural uniformity is characteristic of the association (Beadle and Costin, 1952).

In recent years the probability of floristic communities having discrete boundaries has also been increasingly questioned. Both the Zurich–Montpellier and Uppsala schools appear to have accepted discrete boundaries in their grouping of species into communities. According to Tuomikoski (1942), the Uppsala school's approach to quadrat survey in the field was initially to divide the area subjectively by eye into "associations" and then record qualititatively the species in each association. In doing so, the physiognomy of the vegetation is taken into consideration either deliberately or unintentionally. The Zurich–Montpellier school does not stratify prior to the systematic collection of quasi-quantitative data, but uses field data from numerous stands to recognize associations and then assumes floristic homogeneity in each. Poore (1955), having adapted the Zurich–Montpellier approach to his study, was able to demonstrate that vegetative units selected subjectively may be placed in groups of similar composition (termed *noda*) and that the proportion of intermediate stands is relatively small. More recently at Montpellier, aerial photographs have been used to delineate homogeneous zones in which photographic characteristics are examined and coded, and the zones are then grouped according to their coding and delineated by *isolines*.

On the other hand, Maycock and Curtis (1960), after a detailed study of boreal deciduous woodlands in Minnesota, favored a "vegetational continuum." They considered that the continuum, which exists in time and space, must defy all attempts to segregate distinct associations. A vegetal continuum has been defined as follows:

> If we have a set of sites such that for every site in the group there is at least one other site with which it has one or more species in common and if this group cannot be divided into two or more groups such that all the sites of any one group have no species in common with those of another, then the vegetation may be said to be continuous.

Whether the floristic classification of vegetation should be based on the dominant species of the tallest stratum depends on the purpose of the study. Obviously, this method is the simplest, but it may provide plant classes too broad for microecological studies. The classification, however, is usually adequate for terrain classification, land use and forest surveys, and the planning of detailed ecological studies. Alternatively, if the classification of vegetation is provided by a combination of dominants in each stratum, this may result in an inconveniently large number of small units for ecological studies (Goodall, 1952). A compromise followed by some ecologists is to use groups of characteristic species to classify the vegetation. The concept of characteristic species was introduced by Gradman (1909), expanded by Braun-Blanquet, and expanded quantitatively and statistically using floristic criteria by Goodall (1952).

The species composition of each community is recorded either qualitatively or quantitatively. Quantitative classification of the vegetation using statistical methods is favored by many plant ecologists because it provides objective results, but for practical reasons including cost and time required to complete the survey, complete quantitative floristic surveys usually require to be confined to small floristic areas. Quantitative surveys usually depend on sampling methods that do not cover the entire landscape and, as in the case of forest surveys, may be confined to the woody vegetation or tree stratum only. A floristic quantitative sampling is therefore an approximation to the overall situation, but is far more than a listing of the dominant species recorded along transects.

Two broad methods of floristic description and community identification are recognizable whether this involves classification or ordination. The first method requires a complete listing of all species recorded on the line or in the strip transect or within the square, rectangular, or round sample plot (quadrant) and each species is weighted by a quantitative or qualitative frequency symbol such as abundant, common, occasional, or rare. In the second method the species are simply listed in each quadrant as present or absent and these data, whether based on randomly or systematically distributed plots, can be used to express simply the frequency percentage of each species or may be used to aggregate the species into different communities based on either association (correlation) between the different species or on measures of similarity between the samples.

4.6 VEGETATION AND SOILS

Vegetation is often a good indicator of soil and climatic conditions. Both natural and cultural vegetation types are associated with major climatic regions and characteristic associations of certain types of vegetation with various conditions of local soil and microclimate. Although vegetation is a major indicator of edaphic conditions, its reliability varies from place to place, depending upon other local factors. In some cases, the presence or absence of certain plant structures or plant species will definitely indicate the texture and permeability of the soil, degree of mineralization, nutrient and salinity levels, and so on. For example, a narrow seepage zone may be detected by slight changes in vegetal density and the height of dominant woody species. On the other hand, vegetal cover in high rainfall areas may be dense and uniform, and the type of vegetation may not give any important clue to the nature of the soil or parent material.

As a general rule, moisture, of all the soil conditions, is greatly influenced by topography and is often the dominating influence on the distribution of plant species. However, at times other factors will dominate, and this is an area where much systematic research is needed. In general, the finer the textures, the better the inherent fertility but the poorer the drainage and the longer the time required for the salts to be removed by leaching. Characteristically, therefore, an inland raised beach will show a series of vegetal types, the innermost being on stony and infertile soils, paralleled by others with soils of increasingly fine texture and plant communities richer in plant species.

Because of their control over moisture conditions, geomorphic boundaries that coincide with soil boundaries often determine those of major plant communities. Again, where the bedrock type is expressive of the surface landform, then there is often a conspicuous correlation between vegetation type, soil type, and bedrock. On the limestone and chalk formations of southern England, the floristic composition is distinct from that of plant communities on sandstone and shale and can be related to soil series. In east Africa the plant subformations of the savanna woodland formation are closely associated with the illuvial and eluvial soils. Thus in western Tanzania the valley bottom, black cotton soils (mbuga), are dominated by grassland and shrubland (*Acacia, Combretum*) whereas the red earths (kikungu) and sandy soils (ichenga) of the hill slopes are dominated by woodland (*Brachystegia spiciformis, Julbernardia globiflora*).

Probably the best-known practical example of using the association existing between soils and vegetation is that of the California vegetation-soil survey. First the land is mapped according to its major vegetation cover classes as observed in stereoscopic pairs of aerial photographs and then in the field according to the dominant plant species (Gardner and Wieslander, 1957). Finally, soil boundaries are mapped in the course of detailed field investigations in order to classify the soils to obtain additional information on the floristic composition of the plant communities and to collect information for land management.

FIVE

THE LAND UNIT APPROACH

5.1 INTRODUCTION

The land unit approach to studying and mapping the landscape combines aspects of geomorphology and plant ecology-plant geography. The term *land unit* is regarded as a general term to be used when referring to landscape units of any size (Howard and Mitchell, 1980). It will not be used in the more restricted sense of identifying a distinctive type of unit. It may be noted that remotely sensed imagery from aircraft and satellites is important in the recognition and delineation of such units, because these usually provide the most efficient and sometimes the only practical synoptic base, particularly if topographic maps or appropriate thematic maps are not available.

The clearest and simplest basis is a classification of landscapes into homogeneous units suitable to the mapping scale required for the particular purpose and using essentially landforms and vegetation as the principal criteria for identifying and classifying the units. The land unit approach is especially valuable for reconnaissance surveys. Although it cannot replace separate soil, vegetation, or even geomorphological surveys with their specialized classification units, it can provide them with a useful correlative framework.

A landscape classification based on a genetic ordering of properties cuts across one that is based on spatial distributions, and necessitates a decision over the relative weight to be given to each. The practical need for simplicity in mapping, the usefulness and availability of remotely sensed imagery in identifying vegetation types and landforms, and the geomorphic importance of spatial relationships between land areas justify an emphasis on distributional criteria for defining land units within the landscape.

Landscapes can be subdivided either artificially or naturally. An artificial subdivision aims to avoid subjective bias by using, for example, arbitrarily chosen grid squares, quadrats, or parametrically defined envelopes. A natural subdivision uses perceived variations of the terrain surfaces and plant community, and can often be usefully combined with an artificial subdivision of included areas whose surface variations are slight or not easily observable.

Natural landscape units representing a synthesis of geomorphic features and vegetation, hereinafter referred to as land units, have advantages. First, terrain to a large extent governs the wider distribution of other landscape factors, which can give it an integrating function in their assessment. Second, it is readily visible, divisible, and comprehensible. Third, variations in the vegetation types often emphasize those in the terrain and enable them to be further subdivided. Despite an element of subjectivity in their definition, land units have been practically useful for many years and have made a rich contribution to lay language in descriptive terms of specific areas at a wide variety of scales, such as, in North America, Great Plains, Appalachia, Fall Line, as well as genetic terms such as mountain, hill, beach, gully, forest, woodland, or grassland. Finally, the general availability of aerial photographs and other remotely sensed imagery favors the use of land units because of the ease with which they can be recognized on it. However, in heavily forested areas of the humid tropics, terrain classification is usually more difficult, but considerable discrimination is possible by using associated vegetal patterns and, if available, sidelooking airborne radar imagery.

Not only are land units the most useful reference basis for landscape information but their classification is more useful in the form of a nested hierarchy. Such hierarchies have been so predominantly the most useful type of classification that they have become the commonest. They give the most readily comprehensible arrangement of data. They can be constructed on the basis of a few characteristics and are easily memorized by keeping the number of individuals in a class to a minimum. Their advantages are so striking that they are employed even when this means that the system of affinities must to some extent be distorted (Sokal and Sneath, 1963). Where clear geographical limits are used and categorical overlapping avoided, a nested hierarchical system for terrain can be devised in which each level has a practical utility and interlocking relationship with neighboring levels. The units can be defined by subdivision of extensive areas or by aggregation from the smallest units.

Whether emphasis is placed on vegetation or on geomorphic features in the definition of land units in specific areas usually depends on the magnitude of the units, the variations in the landscape, and the training and experience of the worker. As mentioned previously, the landforms provide information on the development and present condition of the terrain over a long period of time, and the vegetation is indicative in the short term of existing environmental conditions. In combination they synthesize a number of important land characteristics indicative of the current suitability of the land for specified agricultural or other purposes and possibly its potential suitability after improvements have been carried out. Furthermore, practical experience will often indicate that by commencing with geomorphology these land units can be subdivided further with an emphasis on vegetal differences.

Having examined vegetation in several quite different zones of the world, the authors are also of the opinion that, except for the smallest land units (the land element), vegetal structure and not floristic groupings is adequate. In fact, a floristic classification is far more time consuming for practical application and may be less effective.

5.2 DEVELOPMENT OF HIERARCHICAL CLASSIFICATION OF LAND UNITS

In the United States the challenge of rapid development of new lands stimulated an interest in the recognition of natural regions in the late nineteenth and early twentieth centuries. Joerg (1914) reviewed the various approaches to regional subdivision of the country to that time. Bowman (1914) subdivided the country into *physiographic types* based mainly on topographic configuration, water supply, and climate because of the importance of their relationship to human activity and economic values. Heath (1956) noted that Bowman's ideas influenced many surveys between the wars. In 1916 the Association of American Geographers established a committee under Fenneman to define the physiographic regions of the country. They used *section, order,* and *division* as their higher units (Fenneman 1916, 1928). Veatch (1933) gave these ideas practical form by classifying agricultural lands in Michigan into natural land types based on soil topography, natural drainage, and native vegetation, which he foresightedly claimed would have as permanent a value as would soil or geological maps of the area.

The first detailed pioneering study in Britain was by Bourne (1931), who recognized the need for a classification into natural regions for forestry purposes and advocated the use of aerial photographs. Aided by geological maps, he completed an exercise of mapping regions and sites of an extensive area of southeastern England. He recognized that an association of sites really constituted a distinct landscape region, and that a site is an area that, for all practical purposes, has similar physiography, geology, soil, and edaphic factors. His work, involving land units of three different magnitudes, can be viewed as the forerunner of later hierarchical classifications (Howard and Mitchell, 1980). His region and site equate with the contemporary terms *land system* and *land facet*. The land system is judged now to be the most widely used land unit, although on occasions the term *system* may be misunderstood or abused. In fact, the term is so well entrenched in the literature that it would be difficult to change. A land system is primarily a recurrent pattern of genetically linked land facets having a predominantly uniform geology and geomorphic history. The associated vegetation emphasizes the constituent individual facets and defines more clearly the boundaries of the land system.

A first detailed approach to land system classification, more specifically aiming at reconnaissance surveys of large areas of relatively undeveloped land, was devised in 1946 by the Australian CSIRO Division of Land Research and Regional Survey. They placed the emphasis on the land system and the *land facet*, which was then called the *land unit*. The *simple land system* was described as a group of closely related topographic units, usually small in number, that have arisen as a product of common geomorphic phenomena and the whole can only be properly defined by describing these constituent units. They are appropriate to a mapping scale of between ~ 1 : 1,000,000 and 1 : 100,000, the larger scale being necessary in more complex terrain. The intricate geomorphogenetic combination of two or more simple land systems is a *complex land system* and an equivalent nongeomorphogenetically related combination is a *compound land system*. Christian and Stewart (1968) have described the practical value of land systems and enumerated the uses to which the basic units provided by the concept have been put. The land system has also been evaluated and has formed the basis of practical terrain intelligence in Britain, India, and South Africa (Mitchell, 1973). It is approximately equivalent to, though somewhat smaller in scale than, the *région écologique* in the more ecologically based French system (Long, 1974).

Beckett and Webster (1965) used the term *recurrent land pattern* (RLP) for analogous units in the United Kingdom and recognized a distinction between *RLP local forms* and *RLP abstracts*, but in later work they replaced the term RLP with land system. Land system local forms represented actual blocks of country within each of which common lithology and genesis had given rise to a recurrent pattern of recognizable features. The abstracts are groupings of similar local forms, but defined according to a central concept into which they all fall. For example, in southern England the abstract river valley on clay includes local forms on (a) Oxford clay (Upper Thames and tributaries); (b) Kimmeridge and Gault clays (Vale of White Horse); and (c) Lower Lias clay (Severn Valley). The chalk abstract includes the Chilterns, the Berkshire Downs, and the Lower Chalk as its local forms. On the European continent, the Flanders Coastal Dune Belt, Vimy Ridge, and Beauce or Brie represent local forms of coastal dune, chalk cuesta, and chalk plateau land system abstracts, respectively. Comparable examples in arid parts of the United States are the Mohawk Mountains as a local form of crystalline mountains and the Gila River flood plain as a local form of desert river flood plains.

5.3   MAJOR PHYTOGEOMORPHIC UNITS (MACROUNITS)

To provide an overall correlative framework, five macrounits will be described: the *land zone, land division, land province, land region*, and *land subprovince*. Because these units are large, they have both great internal complexity and often vague boundaries. Their value lies in providing a simplifying framework for correlating and mapping smaller units in analogous but separated areas, in aiding reconnaissance of large tracts by supplying an overall view of genetic relations, and in explaining world patterns of distribution of some environmental factors for nonspecialists. The mutual relations of these units in part of the Middle East are shown in Figure 5.1.

Land zones are the largest and each covers one of the generally recognized major climatic zones of the world such as the humid tropics (within which the tropical rainforests occur), the tropical subhumid zone, the Sahel zone, the hot arid zone, the temperate semiarid zone (with its prairies and steppes), the temperate humid zone, and the arctic and subarctic zones (within which the tundra occurs) (Figure 4.1). Land zones are generally mapped at scales smaller than about 1:15,000,000 and represent the basic stratification of the earth's surface into climatic zones within which physiographic processes are sensibly uniform (cf. Chapter 2). The land zone is too large to be easily recognized on individual frames of earth resources satellites, such as Landsat, and can only be appreciated when these are assembled into a mosaic or by viewing high quality but smaller scale environmental satellite imagery (e.g., Tiros/NOAA). Satellite imagery can often help in refining boundaries shown on existing very small scale maps. Many of the highest soil taxonomic classes such as the largest soil units used by FAO–UNESCO (1974) and the soil orders of the U.S. Department of Agriculture (1976) have a distribution recognizably related to that of land zones because of the importance of the climatic factor in defining both.

The next largest unit, the *land division*, has a gross form expressive of a continental geological/geographical structure suitable to a mapping scale of about 1:5,000,000 to 1:15,000,000 (Brink et al., 1966). It can be identified on satellite imagery by variations in tone and color associated with the patterns of its morphotectonic form and the relatively close fit of the natural vegetation to the boundaries. The uniformity of the vegetation is at the level of plant pan formations and of Long's *zone écologique* (1974). Examples of a land division within the humid temperate zone are the North European Plain, the Appalachian Chain of United States, the Great Dividing Range and its coastal plain in east-

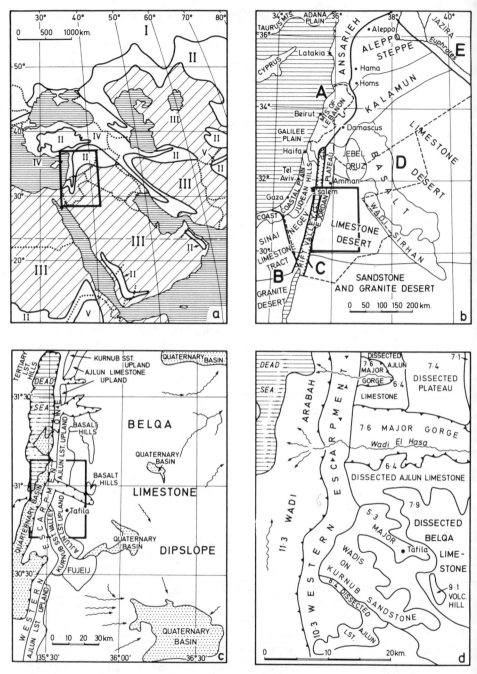

Figure 5.1. Hierarchy of land units of part of the Middle East at a range of scales (*a*) Land zones: I, cool temperate; II, semiarid; III, arid; IV, warm temperate; V, montane. (*b*) Land divisions: A, Levant Coastlands; B, Sinai; C, Rift Valley; D, Arabian Shield; E, Mesopotamia. Land provinces are associated with these: (*c*) land regions and (*d*) land systems. The locations of (*b*), (*c*), and (*d*) are shown by rectangles on (*a*), (*b*), and (*c*), respectively (Howard and Mitchell, 1980).

ern Australia, and in the arid land zone, the Sahara and the Arabian Shield. Because of the very small scale of mapping used for land zones and because delineated boundaries may be based on the scanty data provided by sparsely distributed local climatic stations, it is often more realistic to commence with a phytogeomorphic subdivisive classification at the level of the land division or the land province. Macroclimatic variations are also imbued in the vegetal boundaries of these units, which can be viewed directly on satellite imagery. At this level and at that of the land province and the land region, it is quite common for several main soil units (FAO–UNESCO, 1974) or soil orders (U.S. Department of Agriculture, 1976) to be found in a relatively small area if the region is geologically or geomorphologically complex and of a variable climate.

The *land province* is a subdivision of the land division and can be defined as an assemblage of surface forms and other surface features on a scale expressive of a second-order geological structure or of a large lithological association. Its recognition is often assisted by coincident boundaries of plant formations (savanna woodland, sclerophyllous forest, prairies) that are the equivalent vegetal units. These are suited to a mapping scale of about 1 : 1,000,000 to 1 : 5,000,000 and are readily recognizable on Landsat scenes. Examples of land provinces within the areas mentioned above would be Lowland Britain, the Pyrenees, peninsular Florida, the Great Dividing Range, and the Jordan–Arabah Great Rift Valley. There is some evidence that this unit explains a higher proportion of the variance of land attributes than any other hierarchical level except that of the land facet (Mitchell et al., 1979). It can thus be viewed as next in importance after the land system and land facet and can provide the basis to a simplified subdivisive classification.

The *land region* was defined by Brink et al. (1966) as having a small range of surface form and properties expressive of a lithological unit or a close lithological association having everywhere undergone comparable geomorphic evolution. It is appropriate for a mapping scale of about 1 : 200,000 to 1 : 3,000,000. Examples of land regions include: in Lowland Britain: the Weald, the Fens, and the Hampshire Basin; in the United States: the Cape Cod peninsula; in southern Australia: the igneous intrusions and Silurian Devonian sediments of the western extremity of the Great Dividing Range; and in Jordan: the Aqaba Granite Complex in the desert zone or the Belqa Limestone area of the steppe zone. Because the land region is defined in geological terms, it is difficult to delimit on satellite imagery unless either geological maps or equivalent ground information is available. If these are not available, the imagery analyst may have to carry out a fairly detailed field survey or bypass the land region and subdivide the land province into land subprovinces that are often the easiest units to map on Landsat imagery (Howard, 1976).

Land subprovinces are basically subdivisive mapping units comprising a group of land systems that correspond to Christian and Stewart's (1968) synthesized compound or complex land systems and that have important attributes in common. They appear to correspond to the division of extensive areas of the earth's surface into the ecological life zones of Holdridge (1947), which are based on temperature, precipitation, and potential evapotranspiration and have been widely used in 17 Latin American countries and often equate with plant formations in the Andean region. At the mapping scale and resolution of Landsat imagery (1:200,000–1:1,000,000) a close relation usually exists between the boundaries of the land subprovinces and vegetal boundaries at the level of major plant formations.

The evolving subdivisive classification of the earth's surface into smaller and smaller land units, which are more and more homogeneous, are summarized in Table 5.1. In addition, this table provides a bridge between the macrounits presented in this chapter and the microunits presented in the next.

Whether the land system should be considered as the smallest of the macrounits or the largest of the microunits is a somewhat arbitrary decision. The writers prefer to consider the land system as the major microunit, because in practice this unit, covering ground areas between about 50 and 5000 km^2, is the

Table 5.1.   Example of Hierarchical Land Units in the United Kingdom.

Location	Macrounits	Microunits
Cool temperate	Land zone	
North European plain including Lowland Britain	Land division	
Lowland Britain	Land province	
Midland Scarplands; East Anglia	Land region ≃ Land subprovince	
Chiltern cuesta; Fenlands		Land system (simple)
Chalk escarpment; Shell Marsh-Upper Fen Peat Association in area north of Redmere, Cambridgeshire (Seale, 1975)		Land catena
Moderate chalk slope with grassland or with beech woodland; Upper Fen Peat		Land facet
Complex of silty palaeochannels (rodhams) interdigitating with Fen peat areas		Land clump
Banks of palaeochannels; peat-filled hollows		Land element

(Land facet and Land clump are bracketed together as "Equal scale".)

largest unit important to local land-use planning and land-use management and can be mapped conveniently at scales of 1:50,000 or 1:100,000, which are often favored for national topographic mapping. In practice, land systems can normally be identified relatively easily from the topography and drainage pattern depicted on remotely sensed imagery, and their boundaries will be found to coincide with major changes in the vegetation types and land use patterns.

Table 5.1 also draws attention to the importance of climate as imbued in the land zone as the basis of initially dividing the earth's surface into major units. With the increasing availability of satellite imagery, the boundaries of the land division, as reflecting the interaction of macroclimate and physiography on regional vegetation (plant panformations), can be often traced directly onto the satellite imagery.

At the lower level of macrounits as shown in Table 5.1, there is a choice of proceeding in the subdivisive hierarchical classification of the earth's surface either from the land province to the land subprovince or through the land region before considering the land systems. The reason for this is twofold. First, as already commented on in the text, the land region, which is basically derived from structural geology, may not be readily discerned on the earth's surface, particularly when the work relies only on the analysis of aerial photographs or satellite imagery. Second, as pointed out previously, the land subprovince is readily observed on satellite imagery and is essentially a synthesis of compound or complex land systems, which in itself imbues the interaction of climate and surface geology.

SIX

SUBDIVISIONS OF THE LAND SYSTEM

6.1 INTRODUCTION

There is considerable evidence that the smaller units of the landscape hierarchy are sufficiently homogeneous to justify their wider use. For example, Mitchell et al. (1979) have shown that about two-thirds of the variances of major soil attributes in a single land zone (hot deserts) were explained by partitioning them into the following hierarchy of land units: land division, land province, land region, land system, and land facet, and about 50% of this two-thirds was contributed by the land facet alone. As pointed out by Beckett and Webster (1965) the homogeneity of the facet is maximized by closely specifying its definitive criteria. Furthermore, Webster and Wong (1969) have shown the good agreement between workers with different backgrounds in delimiting land facets (hereinafter in this chapter referred to simply as facets) that closely accord with independently determined soil boundaries in the Oxford area. Usually, larger units do not have a comparable degree of internal homogeneity or reproducibility as recorded by different workers.

Soil units by contrast are defined in terms of profile morphology, and thus are usually more exactly defined and homogeneous, but show more frequent geographical discontinuities between units at the higher categoric levels and thus cannot be directly described by remote sensing data. Although this limits direct comparison between specific examples, comparable hierarchical classifications with a similar number of groupings and some equivalence of categoric levels have evolved.

The land zone, land division, land province, land region, and land system approximately correspond categorically, but not geographically, to the U.S. Department of Agriculture's soil order, suborder, great group, subgroup, and family, respectively (1976), and the land division and land province correspond approximately in the same way to the FAO-UNESCO soil groups and soil units, respectively (1974). The facet corresponds conceptually to the soil series, but geographically it includes at least one and possibly two or three soil series whose boundaries will often not transgress those of the facet. The land catena correlates with the soil association, the land clump with the soil complex, and the subfacet with the soil type. In some areas, notably where soil units have a physiographic expression, land clumps and land elements will coincide with soil complexes and soil series.

6.2 MINOR LAND UNITS (MICROUNITS)

These units are the land facet, the subfacet, and the land element. The facet, like the land system, has been recognized as of major importance and has been defined by a number of workers, and with its equivalents is probably the second most commonly used unit because of its appropriate size for many types of land-use planning and its internal homogeneity as demonstrated in Britain and arid areas (Mitchell, 1973). It is defined by its geology, water regime, topography, and its vegetal structure, but in a much more restricted way than the parent land system. Its terrain features are simple, with a generally homogeneous rock substrate, a single water regime (including both surface and ground water), and a predictable "sense" of internal variation, such that a pedologist would map its soils as an association of soil series, agricultural activities would be uniform, and an engineer would accept a single design specification for a road built on it.

The facet is broadly equivalent to Bourne's *site* (1931) and Holdridge's *association* (1947; Holdridge and Toshi, 1972). Hills (1961) in Canada defined its equivalent, the *physiographic site type*, as a unit where all the "gradients" (variable attributes) important to biological productivity are used in subdividing the next higher unit. Facets vary in size from less than 1 ha to well over 1 km^2, but are appropriate for mapping at scales between about 1:10,000 and 1:100,000, with the larger scales being appropriate for humid and densely populated areas, and the smaller for arid or undeveloped areas.

Examples of facets relevant to any climatic zone are limestone escarpments, small river terraces, alluvial fans, or small dunes. Each land system is made up of a repetition of the facets it contains. The associated vegetation type enables an experienced plant ecologist to identify the facet and its boundaries. The physiognomy of natural vegetation is especially useful in recognizing facets and, with the aid of aerial photographs, it is possible to map quickly the facets within the land system. Except in arid areas or on fallow land, the soil is usually not conspicuous on the photographs and hence cannot easily be used to provide the boundaries of the facets. Although minor changes of slope may occur within the facet, its natural vegetation is usually uniform structurally at the plant subformation level into which it can be subdivided on aerial photographs (Howard, 1970c). It is equivalent to the *station écologique* of Long (1974). In areas where the natural vegetation has been replaced by induced vegetation, the crop patterns and crop types often clearly define on the aerial photographs the boundaries of the land system and frequently the individual facets.

The two units smaller than the facet that have been recognized are the land subfacet and the land element. These have hardly ever been used as mapping units and their interrelationships have been little explored.

Subfacets were first defined as the constituent parts of the facet between which the graduation in properties derives from the dominant genetic processes (Beckett and Webster, 1965). They can best be thought of as parts of the facet where the relatively uniform processes that form the whole facet give rise to distinguishable subdivisions on the basis of material or surface form. If an alluvial fan under arid conditions, for instance, is considered as a facet, the upper stony part where gullies are deeper and the lower more gently sloping finer textured part where they are shallower could be separated as the subfacets. The unit can be practically important if, for instance, a hillslope facet consists of two subfacets—a steep upper and a gentler lower slope, which have markedly different vegetal structures and development potential. Differences in the dominant plant species may also be definitive.

A need has frequently been expressed for a smallest unit that would accord with Wooldridge's "morphological electron" (1932) and would be in Linton's phrase "indivisible on the basis of form" (1951). Christian and Stewart (1968) recognized the need for such a unit, which they called a *site*, which is equivalent to Grant's *component* (1968) and to Brink et al.'s *land element* (1966), the last defined as "for all practical purposes uniform in lithology, form, soil, and vegetation." This is the definition adopted here. It implies uniformity of topography and of vegetation structure at the level of a plant subformation, and uniformity of floristics at the level of the plant consociation or plant association, and is equivalent to Long's *élément de station écologique* (1974).

Examples of land elements would be small rock outcrops and small gullies on hillsides. In the example of alluvial fans quoted to show the nature of subfacets, the land elements would be the gullies, interfluves, nobs, and so on, that are their ultimate landform subdivisions. Each may be expected to be floristically distinctive as a plant community having characteristic dominant species. Because the land element is characterized internally by terrain uniformity, its floristic composition is important in identifying its subdivisions. In regions of the world where there are negligible differences in topography, land elements may extend over considerable areas.

The distinction between local forms and abstracts described in the previous chapter can be applied to facets as well as to land systems. The former represent locally occurring examples of facets, such as moderate to steep scarp on Upper Chalk, whereas the latter represent its central concept, that is, moderate to steep scarp. Another level of generalization can also be recognized—the *variant*. This represents a type of facet distinguished by particular lithology or by differences in the intensity of genetic processes. Examples of the variants of the facet just quoted are moderate to steep scarp on sandstone or on limestone with their distinctive plant physiognomies or land-use patterns. The variant is thus essentially an intermediate unit between the specific facet local form and the general facet abstract.

Both local forms and variants can be regarded as *cognate* subtypes of the facet abstract if they are relatively predictable parts of the land system local forms of which they form a part, that is, if they derive from mere local differences in the intensity of the genetic processes or from minor differences in secondary physiographic processes. If they are relatively unpredictable, for example, because of the presence of isolated areas of vulcanism or exotic deposits, they are considered *noncognate* local forms and variants of the facet.

So far major emphasis has been placed on two land units—the land system and the facet. The variability of the latter both in size and composition may cause it to lack stability as a readily recognizable and mappable land unit. An intermediate unit may thus be needed. Howard (1970b) in acknowledging this problem introduced the term *land catena* into his work in southeastern Australia, and defined it as a major repetitive component of the land system consisting of a chain of geographically related facets (e. g. valley bottom to hilltop to valley bottom). Land catena may be compared with Milne's soil catena and provides the link, based on local toposequences, between the larger land units, which are usually based primarily on landform, and the smaller units, which are increasingly defined in terms of vegetation (Figure 6.1). A transect of the facets in a land catena would be recorded at right angles to the topographic grain of the country. The land catena generally contains a characteristic soil association, and the vegetation assemblage, consisting of several plant subformations, is equivalent in scale to Long's *secteur écologique* (1974). A land system can also be viewed as an aggregation of the land catenas of similar characteristics.

When the terrain is level, the facet may be extremely extensive and consist of land elements that may be much larger than whole facets in rugged terrain. Under these circumstances the land catena may consist of a sequence of land elements in which it is not always necessary to recognize the facet as intermedi-

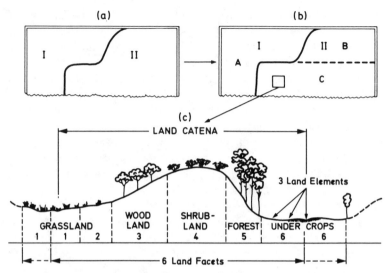

Figure 6.1. Hierarchical subdivision of the landscape. Part of a land province (within double lines) divided into two land subprovinces or land regions (I and II) (*a*), and further subdivided into three land systems (A, B, and C) (*b*). (*c*) A profile (cross section) of a land catena of land system IIC. The land catena is divided into six land facets, primarily because of geomorphic and vegetal variations. Land facet 6 is further subdivided into three land elements primarily because of floristic and pedologic variations, since variation in the topography is minimal.

ate. Land elements may be occasionally so large as to be recognizable on Landsat imagery (e.g., some wadis in the Sudan and some features on the Indus flood plain in Pakistan); whereas it is usually not possible to recognize facets (or even land catenas) in rugged terrain on the imagery because its relatively low resolution and the small surface of each land unit.

Where a facet, for instance, in much dissected terrain, contains a repetition of land elements whose interrelationships are too small or too complex to allow the whole as an acceptable homogeneity, a distinct land unit known as a *land clump* is recognized and can be given equal status in mapping (Mitchell and Perrin, 1966). The relationship between them is therefore analogous to, although usually different in scale from, that between the soil series and the soil complex in soil mapping. Examples of land clumps would be aklé-type dunes or badlands in soft sediments.

6.3 A SIMPLIFIED METHOD OF CLASSIFICATION

Although the authors favor a hierarchical subdivisive classification of the landscape using land units of all magnitudes, they recognize that some workers may wish to use a simplified approach. An example of this, based on critical path analysis, is given in Figure 6.2. Here several land units are omitted for

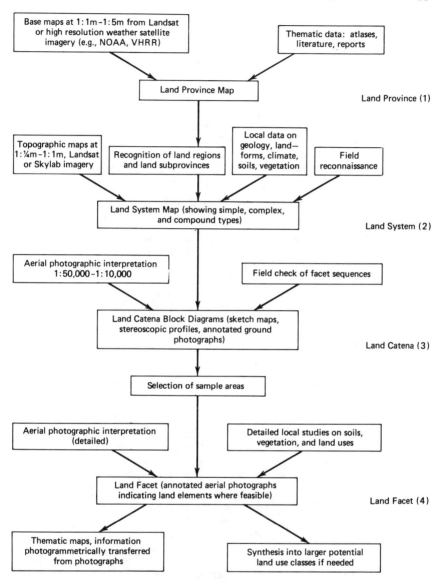

Figure 6.2. Critical path analysis applied to the hierarchical subdivisive classsification of land units (Howard and Mitchell, 1980).

simplicity and to emphasize those of major importance. The procedure is essentially to work from reconnaissance to detail, moving from the identification of the large units on satellite imagery and on aerial photographic mosaics, usually through a process of subsampling of intermediate-sized areas to a selection of critically important smaller units that are recognized and classified

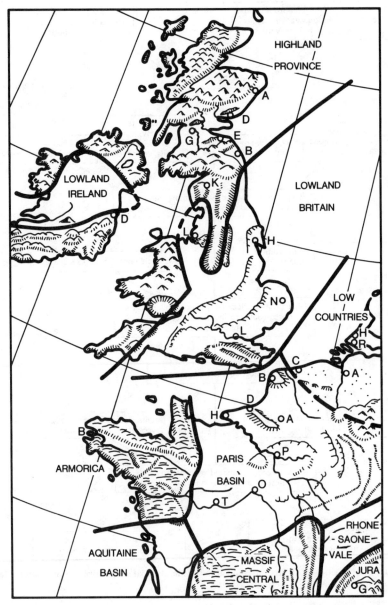

Figure 6.3. Land provinces of a part of western Europe.

with the help of stereoscopic pairs of aerial photographs. These form the basis of local land-use planning.

6.4 MAPPING UNITS

The hierarchical classification of phytogeomorphologically defined land units provides a readily comprehensible, systematic, and practically useful basis for

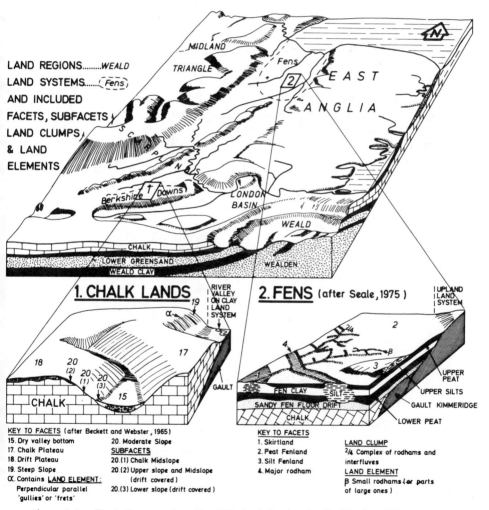

Figure 6.4. Block diagrams of southeast England showing the Chalkland and Fenland land systems and their associated land facets, land subfacets, and land elements.

integrating environmental information whose advantages have been increased by the recognizability of such units on the wide range of remotely sensed imagery now available. Past research and present consensus give a scalar sequence of units recognizable in different parts of the world as follows: land zone, land division, land province, land region, land system, land catena, land clump, facet, subfacet, and land element, each of which has its own rationale and which can be used sequentially in procedures for land resource assessment and for planning purposes.

Several different landscapes in varying climatic zones of the world have been classified according to this hierarchical subdivisive scheme. For such studies both the United States Landsat imagery or the equivalent USSR imagery and stereoscopic pairs of aerial photographs are needed.

A map of part of western Europe (Figure 6.3) in the humid cool temperate land zone illustrates its subdivision into land provinces and, with the aid of geological and topographic maps, provides a basis for further subdivision into land regions and some of their included land systems such as the English Scarplands and Fenlands. These land systems in turn contain an assemblage of smaller units, down to land elements such as the small silty paleochannels standing slightly above the surrounding fen peat (Figure 6.4 and Table 5.1).

SEVEN

REMOTE SENSING TECHNOLOGY

7.1 INTRODUCTION

Remote sensing has an increasingly important role to play in phytogeomorphic studies. The term *remote sensing* was first used in the United States in 1958 in a successful attempt to attract federal funds for research using airborne techniques. By 1961 the term was appearing in the literature published in the United States and was defined as the collection of data about phenomena or objects on or near the earth's surface by a device not in contact with the phenomena or object under investigation. The definition is extremely wide and can be applied to reading this text (human eyes as the sensor), aerial reconnaissance of livestock where observers in low-flying aircraft use their eyes to observe and hence to count objects (the cattle), and the use of spectral radiometers on the ground. Also, the term had been used in a more restricted sense of being confined to the electromagnetic spectrum and hence to exclude shipborne sonar (e.g., for location of fish schools); and has been extended widely to include the analysis of the collected data, whether this is by visual photographic interpretation, or by computer-assisted digital analysis, and the presentation of the analyzed data as information in map form or as statistics.

The definition of the term remote sensing must include the analysis of the collected data, so essential in its application, and recognition of the fact that the platform can be an aircraft, spacecraft, balloon, or ship, and extends to ground-based techniques. With this approach, the term and the French and Spanish equivalents (teledetection, telepercepcion), include aerial photography, satellite sensing, hydrographic survey, aerial and satellite photographic interpretation, aerial and terrestrial photogrammetry, computer-assisted digital analysis of airborne and satellite-collected data, and the presentation of the data as information in the form of planimetric, topographic, and thematic maps, or as statistics. Normally the sensor-collected data can be processed to imagery of the earth's surface (e.g., aerial photographs, Landsat imagery), and hence geophysical techniques that use magnetometers, gravimeters, scintillation counters, and so on, are excluded.

Remote sensing systems of practical application to phytogeomorphology at the present time involve aerial photography using several film–filter combinations and a range of scales, side-looking airborne radar (SLAR), earth resources satellite imagery, particularly the U.S. Landsat, and occasionally environmental satellite imagery (e.g., NOAA AVHRR). The extraction of information, however, still greatly depends on visual interpretation by trained photo interpreters, although the importance of computer-assisted analysis is increasing, particularly in order to provide the photo interpreter with the best quality imagery.

Airborne thermal sensing, although important in pollution studies, has remained very much experimental in studies related to vegetation and landforms; and in general, aerial photography is extensively used and preferred to optical mechanical scanning in the solar spectrum. This is partly due to the resolution of aerial photographs being three to five times finer than scanner imagery taken at the same flying height. In satellite imagery studies, optical mechanical scanning is of major importance because of the Landsat program's use of multispectral scanners (MSS, and TM). Thermal sensing is receiving increasing attention with the availability of imagery from NOAA AVHRR and of the ESA Meteosat.

7.2 COMMENTS ON THE BASIC PRINCIPLES
OF REMOTE SENSING

Because it is impossible in this text to provide a comprehensive review of basic principles, the reader is advised to consult appropriate books on the subject to gain a background to the biophysical properties of objects being sensed, to understand the influence of the troposphere on remote sensing and the characteristics of the solar spectrum, to understand the characteristics and function of remote sensors, the ways in which visual interpretation and digital analysis can be used to derive useful information from the imagery or digital tapes, and the presentation of these data as best suited to the user. A good beginning is to consult the increasing number of books written on remote sensing, or to begin by referring to the 1982 edition of the *Manual of Remote Sensing* published by the American Society of Photogrammetry, or to refer to the several journals of remote sensing, including *Photogrammetric Engineering and Remote Sensing*, *Photogrammetria*, the *International Journal of Remote Sensing*, the *Photogrammetric Record*, and the annual *Proceedings of the Symposia on Remote Sensing of Environment* held under the auspices of the Environmental Research Institute of Michigan (ERIM).

Remote sensing involves the use of passive and active sensing systems. With the exception of recording thermal emitted energy, the passive systems are concerned with the study of recorded reflected energy. Of the imagery derived by active systems, only radar is operational; and the availability of this type of imagery is geographically quite limited, being confined almost entirely to side-looking airborne radar (SLAR). It seems unlikely that radar taken by satellite will be available on a continuing basis until the late 1980s, although synthetic aperture radar (SAR) imagery of the aborted 1979 Seasat program is available over a period of a few weeks for parts of North America, western Europe, and northern Africa.

Side-looking radar, as an *active system*, implies that the energy used to derive the imagery of ground objects is generated in the sensor. This active energy is beamed to the earth's surface, reflected according to the wavelength by the mixture of the vegetation, landform, soil, and so on, and partly recaptured in the sensor.

All other systems of practical application to phytogeomorphology come under the category of *passive sensing*, which relies on reflected solar radiation or on thermal energy emitted by the earth. Incoming solar radiation is in the form of electromagnetic energy, which moves with the constant velocity and harmonic wave pattern of light. This solar radiation at the earth's surface is greatly modified by the general atmospheric conditions including the so-called *atmospheric windows*. These windows are spectral bands through which solar energy passes relatively freely, greatly influencing the spectral distribution of the solar energy received at the earth's surface.

Approximately half of the solar radiation incident on the earth's surface (~ 0.33–3.0 μm) is in the visible part of the solar spectrum (sunlight, ~ 0.38–0.75 μm) and nearly all of the remainder is in the near and mid-infrared. Only a small fraction is in the ultraviolet below 0.40 μm. The boundary between the visible and near-infrared part of the spectrum is variously placed between about 0.725 and 0.825 μm according to the limit of human vision. In the infrared part of the solar spectrum beyond about 0.75 μm, water vapor or to a lesser degree carbon dioxide absorbs much of the incoming solar radiation between the atmospheric windows, and in the far infrared the effective thermal sensing of emitted energy from the earth's surface is confined to the two windows between approximately 4.5 and 6.5 μm and 9.5 and 12.5 μm.

These constraints on passive sensing, combined with the exclusion by glass of all radiation shorter than about 0.38 μm, and the end of the spectral range of sensitivity of photographic film at about 1.0 μm, limit the spectral range of aerial photography. Also, as photographs cannot be taken beyond about 1 μm, a multispectral scanner has to be used at longer wavelengths to capture the reflected solar infrared energy; hence a thermal scanner, possibly as part of a multispectral scanner, is also needed to record thermally emitted radiation at the earth's surface in the 4.5–6.5 and 9.5–12.5 μm bands. There is negligible reflected solar energy above 3.0 μm and very little thermal energy below this value. Nearly all phytogeomorphic surfaces are diffuse reflectors and not specular reflectors. Calm water with its mirrorlike reflective characteristic is an exception.

7.3 SATELLITE REMOTE SENSING

7.3.1 Environmental Satellites

Satellite sensing can be conveniently classed according to whether the satellites are environmental (NOAA, Meteosat, Meteor) or earth resources (Landsat, SPOT, Salyut/Soyuz), and the former can be subdivided according to whether they are *geostationary* (stationary in space relative to the earth's surface) or *earth orbiting*.

Environmental satellites, commencing in 1960 with the polar-orbiting U.S. weather satellite Tiros series, have a very low ground resolution, although as a group these satellites have a high temporal resolution. The European Meteosat-2, a part of the planned global network of meteorological satellites, has a high temporal resolution by providing imagery of Africa and Europe every 30 min, but has a ground resolution of about 2.5 km in the visible spectrum and 5 km in the infrared. The U.S. NOAA/Tiros N series (NOAA 8), with its advanced very high resolution radiometry (AVHRR), provides imagery with a resolution of 1 km in the visible and near-infrared bands and 2.5 km in the two thermal infrared bands. Environmental satellite imagery of this type can be used experimentally to monitor major seasonal changes in vegetation cover over extensive arid and semiarid areas, for desert locust surveillance in Africa and the Near East, and for rangeland management, but provides little new information on landforms.

7.3.2 Earth Resources Satellites

The application of satellite remote sensing to the resources of the earth received its first impetus during the U.S. manned space programmes of the mid-1960s (Mercury, Gemini, Apollo). For example, Mission GT-4 of the Gemini program provided high quality 70-mm normal color photographs at a nominal scale of 1:2,400,000 for geomorphic research and also provided useful information on the distribution of major patterns of plant formations. Other experimental programs providing high quality color photographs of the earth's surface include U.S. Skylab (1976) and the continuing USSR Salyut/Soyuz program. The metric camera in the ESA/U.S. Spacelab mission 1 has provided black-and-white photographs suited to thematic mapping at 1:50,000 and 1:100,000.

7.3.3 Landsat

The U.S. manned space flight photography in the mid-1960s was followed by the highly successful unmanned Landsat program, which commenced with the launching of the Landsat 1 (ERTS-1) in July 1972. This was designed as an experimental system to test the feasibility of collecting earth resources data by unmanned satellites and is continuing through the mid-1980s with the Landsats 4 and 5. Landsats 1 and 2 were each launched with two onboard imaging systems: (a) three-channel return-beam vidicon (RBV) in the visible spectrum functioning similarly to a normal three-camera television system and (b) a four-channel multispectral scanner (MSS) functioning in the visible and near-infrared spectrum (0.5-1.1 μms). The latter functioned similarly to airborne multispectral channel system except that the digital data are transmitted directly to ground receiving stations covering much of the world except for Central America, the Middle East, most of the USSR, and about 70% of Africa. The recording range was about 2400 km for Landsats 1-3 and about 2000 km for Landsat 4. Onboard tape recorders also provided delayed transmission to U.S. ground stations, but the onboard tape recorders of Landsats 1-3 proved erratic in operation.

Relatively few RBV scenes have been obtained, although the same ground area was passed every 18 days and every 16 days with Landsat 5. There are some cloud-prone areas still not adequately covered by RBV or MSS, particularly in the tropical rainforest regions where there is nearly continuous cloud cover and there are no direct readout ground receiving stations. Each MSS scene of Landsats 1-3 covered a swath of about 185 km and a ground area of about 34,000 km^2 with a ground resolution of 57 $\times$ 79 m (0.45 km). In analysis, the smallest phytogeomorphic unit consistently identifiable in the imagery is likely to correspond to 5-10 ha by visual interpretation. The resolution of the

RBV of Landsat 3 was improved to about 30 m, but the newly introduced thermal infrared channel with a resolution of 120 m on Landsat 4 functioned only for a very short time.

Landsat 5 carries a similar MSS system, but no thermal MSS channel and no RBV system. The system was designed to relay directly to the United States via a direct readout telecommunications satellite system (TDRS). However, Landsat 5 has on board an advanced multispectral scanner termed the *Thematic Mapper* (*TM*) that operates in seven well-chosen channels suited to mapping vegetation including agricultural crops. Bands 1–6 have a resolution of about 30 m (0.45–2.25 μms), and band 7 in the thermal infrared range of 10.50–12.50 μms has a resolution of 120 m. The USSR Salyut/Soyuz imagery and that of the French SPOT satellite have a ground resolution of 10–20 m. The SPOT will also provide stereoscopic imagery and its side-looking capacity will provide imagery of the same ground area every 5–6 days. The purchase price of the imagery is likely to be several times that of older Landsat types of imagery.

Landsat imagery and tapes have had the advantage of being cheap to purchase, because the price reflects only the cost of reproduction of the imagery or tapes and management costs, and does not include development costs, launch costs, and cost of the satellite. One Landsat scene would require about 1600 aerial photographs at 1:40,000 scale to cover the same ground area and the cost of the aerial photography would be several hundred times greater. The resolution of the Landsat MSS imagery is, however, more than 40 times lower than that of the aerial photography, but at the same time TM approximates the resolution of most real aperture side-looking airborne radar (SLAR) imagery, the taking of which is likely to be comparable in cost with small-scale aerial photography.

7.4   AERIAL PHOTOGRAPHY

The taking of aerial photographs is the oldest airborne remote sensing method used in phytogeomorphic studies. Modern vertical aerial photography can be viewed as beginning with the preparation of a geological map in 1911 of the landscape near Benghazi in Libya (Tardivo, 1913). This was followed by a forest-type map in eastern Canada in 1918 (Wilson, 1920) and tropical vegetation mapped in Burma in 1922–1923 (Kemp et al., 1925). Bourne (1931) in his studies of land units near Oxford and later studies in Zambia and Burma examined the landscape using black-and-white panchromatic aerial photographs to identify land units using both vegetation types and the principal landforms; hence this can be viewed as the first phytogeomorphic study with remote sensing data. Considerable progress was also made in applying aerial photography to landscape evaluation in the Pacific area during World War II (Belcher et al., 1951) and this was followed from 1945 onward in northern Australia by the use by CSIRO of small-scale black-and-white aerial photographs in land system surveys (Christian and Stewart, 1968).

In the past 20 years the versatility of aerial photography has increased greatly using different film–filter combinations and focal length lenses. Flying heights range from about 100 m using a light high-wing aircraft to about 13,000 m with civil aircraft and 20,000 m using military reconnaissance aircraft with corresponding photographic scales between 1:1,000 and 1:150,000. However, the scale of readily available photographs is most likely to be between about 1:10,000 and 1:85,000. Conventional aerial photography with modern cameras (Wild RC-10) and fine-grain film provide imagery with a ground resolution of less than 1 m. High altitude photography can provide a ground area coverage of more than 10,000 km^2 a day and imagery with a ground resolution of finer than 2 m. The information content is better than that in older black-and-white photography at 1:30,000. In phytogeomorphic studies scales between about 1:15,000 and 1:60,000 may be preferred because there has to be a compromise between scales best suited to vegetation studies and those best suited to geomorphic studies. Larger scales are usually preferred by the plant ecologist and forester, because this enables vegetation structure to be examined, some tree and shrub species to be identified, and estimates to be made of several stand parameters, including stand height, crown cover percentage, and crown diameter. Photographs at smaller scales are usually preferred by geomorphologists, because they prefer to view comparatively large areas of the landscape and their examination of the landforms, including drainage patterns and topography, is best not distracted by minor changes in the patterns of the vegetation.

There is the broad choice among black-and-white panchromatic film, which remains the most popular, infrared black-and-white film of somewhat lower resolution but with superior haze-penetrating capability, infrared color film with similar haze-penetrating capability but with the advantage of providing a color record of the landscape, and normal color, which is becoming increasingly popular provided the local conditions are not exacting on the taking of photographs. Exacting conditions include strong contrasts between light and dark areas such as the illuminated and shaded sides of mountain ranges, low sun angle and atmospheric haze produced by dust, and the burning of extensive areas of vegetation. Infrared photography usually provides more information on drainage patterns and a clear discrimination between wet and dry exposed surfaces. Some vegetation types are easily separated. Indigenous hardwoods can be distinguished from conifers in the northern hemisphere, and major changes in the density of the vegetal cover usually show up on color infrared photographs. However, shadows are enhanced more by infrared photography than by panchromatic photography, which results in loss of detail in shaded valleys and under trees; and with these conditions the best results may often be obtained using the normal film–filter combination of black-and-white panchromatic film with a minus-blue filter.

7.5   SIDE-LOOKING AIRBORNE RADAR

The acronym *radar* is derived from radio detection and ranging and was first widely used in World War II for the detection of hostile ships and aircraft and for navigation. In the 1950s airborne radar imagery became operational for military purposes and side-looking airborne radar (SLAR) became available for nonmilitary purposes in the 1960s. The first large-scale successful SLAR project was the thematic mapping in 1968 of part of Panama (20,000 km^2) using K-band SLAR.

Side-looking airborne radar depends on a sideways-facing slit antenna mounted below the aircraft, hence the prefix side-looking. The quality of the imagery is usually improved by increasing the length of the slit antenna. This slit, which is fixed and does not mechanically scan as compared with thermal sensing systems, actively emits a narrow spreading path of radio waves of known wavelength (e.g., X-band) at right angles to the aircraft, known as the range direction, and scans in a forward direction due to the forward movement of the aircraft, known as the azimuth direction. The time for the return of the signals from the ground objects depends on the slant range distance between ground and aircraft, and the returning signals are recorded in the aircraft directly on tape or film that is moving forward at a rate proportional to the ground speed of the aircraft. In order to be imaged separately, two ground features that are close together must reflect differently in time so that their signals are received separately by the antenna on the aircraft.

To reduce the antenna length and at the same time to improve the resolution of the real aperture radar, also called noncoherent or brute force radar, synthetic aperture radar (SAR) has been developed. Synthetic aperture systems use a very short physical antenna; but by detecting what is termed *Doppler frequency shifts*, which are changes in the wave frequency as a function of the relative velocities of a transmitter, the effect of a very long real aperture antenna is synthesized. For example, a synthetic aperture antenna of 2 m on an aircraft can effectively function equivalent to a real aperture antenna of about 600 m in length. Synthetic aperture radar was used in the aborted Seasat satellite program in the 1970s and is part of a cooperative western European program in the late 1980s.

In interpreting SLAR imagery, there are three major problems that arise from the fact that the imagery is formed from the effect of a diagonally ranging energy source emitted and collected by the antenna under the aircraft. First, imagery cannot be obtained of the landscape below the aircraft. Second, the resolution of the imagery is coarse as compared with aerial photographs taken at a similar flying height. Individual trees are seldom clearly resolved and smaller vegetation types may be difficult to classify, although some geomorphic features are very well defined (e.g., drainage patterns).

Third, the side of a hill in the ranging direction may be excessively illuminated if inclined toward the aircraft and in total shadow if the inclination is away from it. Hence in rugged terrain little information may be obtained in the "look" direction of the landscape beyond high mountains, although this may be partly overcome at considerable extra expense by sensing the same ground area in two flight directions (see Figure 8.5). In addition, because the image is an oblique record of the landscape, it has a varying geometric scale, but this is now overcome by computer-processing digitally stored raw SLAR data, and not recording it directly on film. On the other hand, radar is virtually free of time constraints because it can be used night or day and in most weather conditions, since it has the capability of penetrating clouds and even light to moderate rain.

7.6 ANALYSIS EQUIPMENT

Comparable development of data analysis equipment has usually followed the development of remote sensing data collecting systems. The motivation has come from the need for mechanical data handling, particularly due to the requirements of optical mechanical scanners (e.g., Landsat MSS), the greatly increased amount of remotely sensed data, and the increasing availability at lower cost of computer-assisted techniques. Twenty years ago methods of automated photogrammetric mapping, including orthophotomapping, were only being developed and tested, digital imagery analysis was unknown, and imagery analysis relied almost entirely on the human visual interpretation of aerial photographs. Unfortunately, at the present time there is a tendency to consider visual photo interpretation as outdated and ready to be replaced by digital analysis using computer techniques; whereas the better approach would be to recognize visual and digital techniques as complementing each other.

7.6.1 Aids to Aerial Photographic Interpretation

Traditionally, visual photographic interpretation as an analogue method begins with the examination in the laboratory and in the field of landforms as depicted on aerial photographs or photographic mosaics and the stereoscopic examination of vegetation using stereoscopic pairs of aerial photographs.

Equipment includes simple lens or pocket stereoscopes, mirror stereoscopes, zoom stereoscopes, polar and electronic planimeters, parallax wedges, parallax bars, engineers scales, and light tables.

The pocket stereoscope has the advantage of very low cost, smallness, and robustness, but the disadvantage is that only a very limited area of the aerial photograph can be observed at any one time.

Because in phytogeomorphic photo interpretation it is usually necessary to commence by viewing as large an area as possible, a mirror stereoscope (with magnifying monoculars for examining the vegetation) is essential. This inexpensive instrument extends the observation rays to about half the area of each photograph with the aid of mirrors and prisms. The stereoscopic model of the landscape from a contiguous pair of aerial photographs observed under the mirror stereoscope conveniently exaggerates the vertical relief of the terrain. If required, height measurements (measurements in the z direction), which include slope percentage, spot ground heights, and tree height, can be made

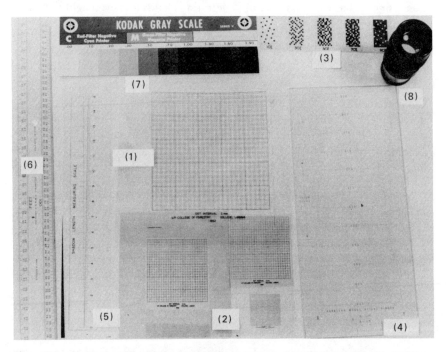

Figure 7.1.   Measuring aids: (1) dot grid with 2 mm between dots; (2) three grids with dot intervals of 1 mm, 0.5 mm, and 0.25 mm; (3) crown closure (crown density) scale; (4) parallax wedge; (5) shadow wedge or crown diameter wedge; (6) engineer's scale graduated in 1/50ths of an inch so that readings can be made to 1/1000th inch; (7) Kodak tonal scale; (8) monocular magnifying scale.

using a parallax bar in combination with stereoscopic viewing. A scanning mirror stereoscope provides further versatility by enabling any common area of contiguous pairs of aerial photographs to be examined stereoscopically without having to realign the stereoscope or to adjust the stereoscopic pair of photographs.

Horizontal measurements (in x and y directions) can be made crudely using a school millimeter ruler or a range of simple measuring aids including a parallax wedge, engineer's scale, and optically magnified scales (Figure 7.1). These are, however, only approximate measurements because the single aerial photograph is a perspective view and scale changes throughout the photograph according to topographic displacement and aerial photographic tilt. Alternatively, horizontal measurements can be obtained more accurately using a stereoscopic tracing stereometer attached to a parallax bar or with a pantograph arm attached to a third-order planimetric or topographic mapping instrument.

The transfer of aerial photographic details onto a map base is achieved best by using a stereoscopic mapping instrument. These include third-order topographic plotters (e.g., Zeiss Stereocord; Figure 7.2) or simple optical transfer instruments (e.g., Multiscope, Bausch and Lomb Transferscope, OMI Stereo-Facet Plotter). Other techniques include the use of single photograph vertical sketchmasters using the camera lucida principle or simply transferring at contact scale onto a transparent overlay the details contained in an assemblage of, say, 4 to 12 aerial photographs (Howard, 1970a).

7.6.2 Other Analogue Aids

The traditional techniques of nonstereoscopic aerial photographic interpretation are commonly used in the study of satellite imagery, but often analysis involves the use of specialized analogue equipment. In phytogeomorphic studies, when using traditional photo interpretative methods with satellite imagery, workers will find that instead of having to be highly selective of the type of information that can be synthesized, as in the case of aerial photo interpretation, they have to make the maximum use of the very limited usable data contained in the satellite imagery.

Modern specialized analogue equipment is designed to take this problem into consideration. This may require the conversion of analogue imagery to a more useful analogue format or sometimes to a digital format (A to D conversion) or the digital format to analogue (D to A conversion). The A to D output is achieved by incorporating a scanning densitometer in the equipment. This provides a record on magnetic tape of the gray-scale density levels of the images as a matrix of fine discrete binary integers or picture elements (pixels). Frequently the discrete gray-scale values are expressed in the range of 1–255, or

Figure 7.2. Several compact monocular or stereoscopic instruments are well suited to phyto-geomorphic mapping. Illustrated is the Zeiss Stereocord, which employs optical–mechanical principles in combination with a desktop calculator or minicomputer to transform coordinates extracted from the stereoscopic pair of photographs on the platen in the center of the figure into ground coordinates. These ground coordinates are used for computation of lengths, areas, distances, heights, and slopes, and, in combination with a pantograph, for graphical mapping.

1–125, or 1–63, whereas the human eye will probably detect no more than about 10 gray levels.

Three types of analogue equipment have been frequently used in phytogeomorphic studies of Landsat imagery: (a) the additive viewer imbuing optical principles; (b) the Diazo printer using photographic techniques; and (c) the density slicer incorporating closed-circuit television.

Modern *additive viewers* were developed initially in the 1960s to be used with multispectral aerial photographs, although the principle of the additive viewer was demonstrated to the Royal Society in London at the end of the nineteenth century. Basically, three black-and-white transparency projectors are used in conjunction with color filters, one for each primary color (blue, green, red). The three projected scenes in primary colors are then combined into a single multicolor picture on a viewing screen. Because Landsat MSS imagery is in four bands, commercially manufactured additive viewers have four projectors with a separate rotating color wheel to provide color control (Figure 7.3).

(a)

(b)

Figure 7.3. Two views of an additive viewer (Clyde Surveys). Each of the 4 simultaneously recorded Landsat bands can be projected onto a screen by light passing through a colored filter, to give a color composite image. (a) Side view, (b) lengthwise view: screen is in center, control buttons to right and left, and projectors behind.

99

A variable resistance on each projector also controls the brightness of each scene. Either positive or negative black-and-white transparencies can be used to provide an enlargement (e.g., 3×) on the viewing screen as a natural or false color imagery. The final hardcopy product is usually a 23 × 23 cm or 46 × 46 cm color enlargement. The advantage of the system is the delicate interactive control by the operator of the hue and brightness of the color composite.

The *Diazo color printer* (Figure 7.4) is simple to operate and much cheaper to buy than an additive viewer. It gives single color transparencies at contact scale, which can be manually combined into a false color scene. Analysis is achieved by viewing this combination of several photographically processed transparencies on a light table. The quality of the imagery is comparable to that of the additive viewer, but viewing is relatively cumbersome as there is no

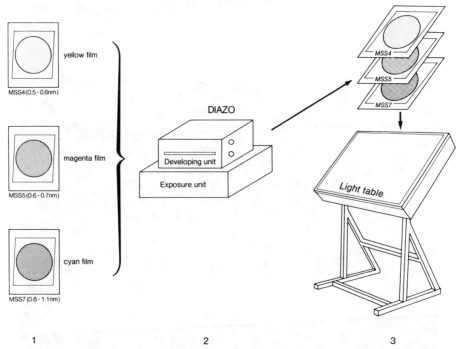

Figure 7.4.   The use of low-cost processing for Landsat interpretation: (1) Three spectral bands from a Landsat frame are exposed onto different-colored diazo films. (2) They are then processed and developed at contact scale. (3) The imagery in the separate bands can then be superimposed and viewed on a light table to give a color composite image. The combination usually preferred is yellow, magenta, and cyan films for diapositives of the 0.5–0.6, 0.7–0.8, and 0.8–1.1 μm bands, respectively (complementary colors to these are used for negative film). This gives a false-color image.

visual control over the hue, value, and chroma of the color composite except by repeated processing.

Different in principle to the two previously mentioned types of equipment is the *density slicer*. This is basically a closed-circuit television system, that is used for viewing the black-and-white single transparencies or prints on a color television screen (Landsat scenes, aerial photographs).

In the process of viewing the raster scanned scene, the video signals are processed to be displayed as color-coded density sliced imagery. Color-coded gray levels on the imagery can be expanded, contracted, or eliminated. The photo interpreter is able to view interactively a much wider range of gray levels than the human eye normally detects in black and white. Depending on the type of density slicer, it may be possible, using floppy disc storage, to combine up to three scenes in primary colors on the television screen, to estimate areas of the same class provided the features have unique gray tones, and to provide a line display of gray-scale densities across the scene. Although cheaper to purchase than digital equipment, it lacks the latter's versatility and speed of operation and has been largely replaced by digital systems.

7.6.3 Digital Analysis

Digital techniques have several advantages over long-established visual interpretation techniques. The latter rely greatly on the skill of the well-trained mind with local experience to evaluate spatial patterns and textures in addition to radiance values in the form of gray tones and colors. Digital analysis is quick in execution, suitable to large amounts of data, is far less labor intensive, because the processing is computer assisted, handles multidimensional spatial data by not being confined to three primary colors and is not limited to two- or three-dimensional display in space. Digital analysis performs well when spectral landscape patterns are dominant and the radiance values of the scene can be represented by discrete numerical classes of individual pixels.

However, there is not only a major increase in capital investment when computer-assisted analysis is used, as mentioned previously, but also considerable increase in operating costs and operating skills. It is recommended therefore to assess the choice of a planned system by a cost–benefit analysis. Frequently technologists fail to recognize the fact that the time and the money spent on designing a remote sensing system, including the identification of the equipment to be used and consideration of the needs of the user community, is relatively a very small part of the total cost.

In its simplest form the hardware of a digital system consists of a tape drive and associated control minicomputer, a host computer, and image disc or magnetic tape drive in which processing occurs, magnetic tape or disc storage, a keyboard to provide stepwise instructions, a television-type color monitor and output devices (photographic film writer, electrostatic black-and-white printer, color hardcopy dot line printer, or flatbed plotter). The color or occasionally black-and-white display usually allows the display on a closed-circuit television screen of 512 × 512 pixels (picture elements), although television screens in Europe provide 625 lines or about 0.7 million discrete pixels, and displays at higher cost can provide 1024 × 1024 pixels. The color display normally allows the viewing in three bands of pixels covering about $\frac{1}{25}$ of a Landsat scene. By viewing only every fifth pixel, the whole scene can be viewed and doubling each pixel (e.g., quadrupling the number of pixels) allows a part of the area to be viewed at double scale. Although one hardcopy Landsat color print (23 cm × 23 cm) may be considered to have about 30 million linearly arranged pixels, a single 35 mm transparency has about $3\frac{1}{2}$ million nonlinearly arranged silver halide elements.

Although in real terms the cost of the hardware has decreased considerably in recent years, the cost of the software programs, which control the operation of the hardware, has not fallen proportionately in price. In fact, the software programs will represent a significant part of the overall cost. A hardware system suited for classroom training with three-channel color display and simple software controls may cost as little as $10,000 in the United States; but a flexible digital system with color viewing of three of four Landsat bands, adequate disc storage, geographic correction, and a menu of statistical analysis software programs may be at least ten times more expensive.

Many of the software programs commercially available in recent years have been developed to handle the overwhelming volume of Landsat data, which as primarily recorded, are in digital form. Each Landsat band is an array of 3240 × 2340 pixels, and each Landsat scene in four bands will thus consist of approximately 30 million pixels.

In conclusion, some of the more commonly used Landsat software techniques will be categorized (see Lillesand and Kiefer, 1979).

First, *image restoration* may be needed so that the hardcopy will be a more faithful representation of the landscape and freed of distortion due to variations in the spacecraft's altitude, velocity, attitude, and so on. However, this computer-based procedure to correct radiometric and geometric difficulties is increasingly being undertaken by the primary data handling organizations (e.g., NOAA, ESA). Nevertheless the imagery analyst may still need software programs to correct the scene to a commonly used map projection (such as Lambert, horizontal Mercator), to correct the radiance values of adjoining scenes when making a mosaic map, or to reduce radiance distortion prior to temporal comparison of pixels of the same ground area.

Second, as a preprocessing operation prior to displaying imagery for visual analysis, or possibly before undertaking computer-assisted classification, software programs are used to enhance the images recorded as pixel values in the Landsat scene. Commonly used *software enhancement programs* include *contrast stretching, two-band ratioing*, and *pixel transformation*. In *contrast* linear stretching, the range of pixels representing ground objects are displayed over a fuller range of gray-scale values, and in *histogram-equalized stretching* more gray-scale values are assigned to the most frequently occurring portions of the histogram. In *two-band image ratioing*, spectral contrast of images is generated by computing digital numbers based on the ratio of the gray-scale values of the image in two separate bands. This has the effect of enhancing spectral differences and, in combination with haze minimization techniques, reduces the effect of differential illumination across the Landsat scene. In pixel transformation, statistical techniques (principal components analysis and canonical analysis) are used to transform the pixel gray-scale numbers into an alternative set of values.

Third, *image classification techniques* may be used to provide automatically interpreted analysis or to help the image analyst to operate interactively with the computer display. In this process each pixel is evaluated within the computer and assigned by it to an information category. These statistical techniques include unsupervised and supervised classification.

In *unsupervised classification*, which may be used prior to supervised classification, advantage is taken of rapid processing capabilities of the computer. The computer, using algorithms, examines the pixels and divides them into artificial groupings based on the pixel values. Advantage is taken of the fact that within the Landsat scene, the pixels representing the radiance values of several major features of the landscape, may tend to cluster naturally into separate classes and each class in the scene is represented by its color-coded values. That is, each pattern class does not have a distinctive radiance value, but is represented by a group of pixels with a recognizable centralizing tendency. Such a procedure is termed *cluster analysis*. Several *supervised classification* strategies use a training set of descriptors that are calculated, examined, and modified by the analyst in order to group unidentified pixels into classes. These include using arithmetic *minimum distances to the means*, minimum and maximum values to provide the *category ranges* (and its extension *parallelipiped classification*), and Gaussian *maximum likelihood* based on the variances of each class, which are assumed to be normally distributed. Further improvement can often be made by combining these classification methods with training sets of pixels carefully identified in the field and/or on single or stereoscopic pairs of aerial photographs.

It is possible to manipulate the pixel values digitally from each of the four spectral bands of the Landsat multispectral scanner in order to maximize all the internal contrasts in the imagery. The method used is *principal compo-*

nents analysis. This gives *eigenvectors* that explain the causes of the *eigenvalues* which quantify the relative explanatory importance of each cause in terms of percentages of the total variation. A principal components transformation and a reconstruction of the imagery in analogue form based on the dominant eigenvectors give optimal resolution of all the internal contrasts of the imagery and is sometimes preferred to the simple combination or ratioing of the original spectral bands.

EIGHT

APPLIED
REMOTE SENSING

8.1 INTRODUCTION

The main justification for integrating remote sensing into phytogeomorphic field studies is to reduce cost and save time. Often field sampling can only be justified as worthwhile because of the cost saving contribution of remote sensing imagery analysis. As long ago as the 1950s, the use of aerial photographs in forest survey in Canada was found to reduce survey cost up to 80 times; and in the same period, the introduction of aerial photography and statistical sampling into forest land system assessment in western Tanzania enabled ten times the ground area to be covered without increasing the size of the field teams (Howard, 1959). In addition, remotely sensed imagery sometimes provides unique data, which would not be available from ground-collected data alone.

The analysis of remotely sensed data should be viewed as complementing essential fieldwork in phytogeomorphology. Remotely sensed imagery permits extrapolation from field observations and ground data serve to check and improve the quality of the imagery analysis. This interaction leads to the development of sampling techniques that combine the greater accuracy of field sampling with the speed provided by analysis of the imagery.

8.2   FIELD SAMPLING

As sampling using remote sensing imagery is far quicker than ground sampling, it is advantageous to sample in the field only a fraction of the plots sampled on aerial photographs or satellite imagery and to collect in the field only the data that cannot be otherwise obtained. This is termed *subsampling*. An initial examination of the imagery will provide field-workers with an overall impression of the distribution of landforms and vegetation types within the area under examination, will help them choose the best routes to key ground areas, and will contribute to the design of the field sampling. For many purposes the approach to using remote sensing can be considered as based on deductive reasoning. That is, when attempting to analyze imagery one proceeds from general observations of the entire scene to localized observations.

Two principal methods of subsampling can be used in phytogeomorphic studies to combine imagery analysis with fieldwork (Figure 8.1). If the size of the individual plot on the imagery is retained, but only a percentage of the plots is sampled in the field, the method has been termed *multiphase sampling*. If each subsample in the field forms but a part of each plot on the remotely sensed imagery, then the term *multistage sampling* is used. In the field the process is illustrated by sampling first the landforms, then within the landforms, sampling small plots to collect data on the vegetation and soils. This provides, in combination with remotely sensed data, triple or quadruple sampling depending on whether satellite imagery and aerial photographs are used.

In Iran, for example, a triple sampling design was used (Rogers, 1961). The triple sample comprised 2381 aerial photographic points for determining the

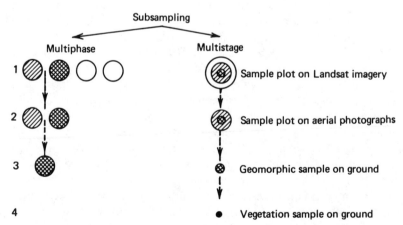

Figure 8.1. Subsampling may be multiphase or multistage, first using satellite imagery, then corroborative data from aerial photographs, and finally the sampling of a much reduced area in the field.

area of forest land and nonforest land. Then 2308 photo plots were used for taking forest stand measurements and finally 54 ground plots were used for obtaining normal forest inventory data. Point sampling of aerial photographs for determining directly by area the landform types and vegetation classes has been termed *photomapping* and has the advantage of not requiring the preparation of a thematic map.

A number of statistical techniques are also useful when comparing imagery data with field data. These include the *t*-test, chi-square test, variance analysis, and regression analysis. Two useful tools of double sampling are the separate calculations of the correlation coefficient and of the regression between data from the photographic sample plots and from the same plots in the field.

The determination of the correlation coefficient between photo plots and field plots is helpful in indicating whether a double-sample survey is likely to be as acceptable as a field survey without photographs. The correlation coefficient provides an estimate of the (linear) association between numerical data derived from the photo plot and from the corresponding field plots. If the association is poor, then the coefficient tends toward zero and conversely if the correlation is high, the value approaches $+1$. Values approaching -1 indicate a strong inverse relationship.

Regression analysis has also received considerable attention in studies involving imagery analysis and field sampling (Rogers, 1961; Stellingwerf, 1963). The regression analysis depends on the recording of measurements, such as stand height, canopy cover, slope, and drainage density, from (a) a relatively large number of photo plots and (b) the corresponding measurements of the same plots in the field for a portion of the photo plots. The measurement of the smaller number of plots in (b) is used in calculating a regression constant and regression coefficient. The commonest form of the regression equation, assuming a linear relationship, is

$$y = a + b(\bar{x}_1 - \bar{x}_2)$$

where $y =$ adjusted mean photo measurement of all photo plots
  $a =$ regression constant; mean of all field plot measurements
  $b =$ regression coefficient; calculated for field on photo measurements
  $\bar{x}_1 =$ unadjusted mean of measurements of all photo plots
  $\bar{x}_2 =$ mean of the measurement of the photo plots also sampled in the field

8.3 ENVIRONMENTAL SATELLITE IMAGERY

As mentioned previously, satellites used primarily to provide meteorological data are now receiving increasing attention in vegetation studies, despite their low ground resolution, because they have a high temporal resolution. In the medium term, the data to be obtained from the higher resolution radiometers on this type of satellite may provide key inputs into multivariate models, which combine spectral and climatic data and which could be used in the global and regional monitoring of vegetation changes (Figure 8.2 a–d; see color insert).

The feasibility of using AVHRR imagery of the NOAA Tiros-N series of satellites has been studied in several countries including the United States (e. g., Texas, Norwine and Greegor, 1983) and Senegal (Tucker et al., 1983). Gray and McCray (1981) were able to show that there is a high correlation for vegetation greenness using NOAA AVHRR data sets and Landsat MSS data sets.

A vegetation greenness index or normalized difference vegetation index (ND) can be devised by ratioing the radiances of the visible and near infrared bands.

$$ND = \frac{NOAA\ AVHRR\ channel\ 2 - channel\ 1}{NOAA\ AVHRR\ channel\ 2 + channel\ 1}$$

Several studies have shown that the currently used remote sensing techniques are not sensitive to nongreen leaf components of phytomass, but are strongly correlated with green leaf area and some types of net production (Norwine and Greegor, 1983). The vegetation index of the Texas study showed the lowest correlation with phytomass ($r = 0.81$), and the highest with leaf area index ($r = 0.91$), and net plant productivity ($g/m^2\ yr - r = 0.95$).

8.4 EARTH RESOURCES SATELLITE IMAGERY

In the past decade analysis has focused on MSS data using analogue and digital techniques. This is primarily because the Landsats are the only family of satellites that have provided continued coverage of most regions of the land surface of the earth.

Landsat imagery should be seen as complementary in use to aerial photographs (and SLAR) and as a powerful tool in multistage sampling for which it may form the first interpretative stage, preferably followed by the interpretation of aerial photographs as the second stage, and then as the third or possibly fourth stage, the collection of data on landforms and vegetation in the field. Although favorable climatic conditions may provide several Landsat scenes of the same ground area every 1–2 yr, it is unlikely that repetitive aerial photographic coverage will be available for less than 5–10 yr and often much longer periods, mainly because of the high cost of aerial photography.

An indication of the relative usefulness of Landsat imagery and aerial photographs is seen through application in land-use studies. In Table 8.1 the land-

Table 8.1   Land Use Classes Suited to Remote Sensing.[a]

Level I		Level II	Level I		Level II
1	Urban or built-up land	11 Residential	5	Water	51 Streams and canals
		12 Commercial and services			52 lakes
		13 Industrial			53 Reservoirs
		14 Transportation, communications, and utilities			54 Bays and estuaries
		15 Industrial and commercial complexes	6	Wetland	61 Forested wetland
		16 Mixed urban or built-up land			62 Nonforested wetland
		17 Other urban or built-up land	7	Barren land	71 Dry salt flats
					72 Beaches
					73 Sandy areas other than beaches
2	Agricultural land	21 Cropland and pasture			74 Bare exposed rock
		22 Orchards, groves, vineyards, nurseries, and ornamental horticultural areas			75 Strip mines, quarries, and gravel pits
		23 Confined feeding operations			76 Transitional areas
		24 Other horiticul-land			77 Mixed barren land
			8	Tundra	81 Shrub and brush tundra
3	Rangeland	31 Herbaceous rangeland			82 Herbaceous tundra
		32 Shrub and brush rangeland			83 Bare ground tundra
		33 Mixed rangeland			84 Wet tundra
4	Forest land	41 Deciduous forest land			85 Mixed tundra
		42 Evergreen forest land	9	Perennial snow or ice	91 Perennial snow fields
		43 Mixed forest land			92 Glaciers

[a]Level I is appropriate for satellite imagery analysis, but level II requires using aerial photographs (after Anderson et al., 1976).

scape at level 1 can be divided into nine classes using Landsat imagery. Furthermore subdivision at level II requires the use of the satellite imagery in combination with at least small-scale aerial photographs and topographic maps. Further subdivision of the landscape into level III (not shown) requires larger scale aerial photographs (1:15,000 to 1:40,000) maps, and supplementary field derived data.

The most appropriate Landsat bands in phytogeomorphic studies are bands 4, 5, and 6 or 4, 5, and 7. Band 4 is often preferred for detecting cultural features and in areas of deep clear water the penetration is often best using band 4. Band 5 usually provides better troposphere penetration, although band 7 is usually better than band 6 in this respect. Bands 6 and 7 are best for discriminating wetlands and water surfaces. The combination of band 5 with band 6 or 7 is usually best in separating areas of vegetation from areas without vegetation and for studies of plant cover. The combination of bands 5 and 7 is also helpful to geomorphic studies in lower rainfall zones.

Most frequently the imagery analyst will commence by examining the black-and-white single band prints of the Landsat imagery at 1:1,000,000 and then proceed to studying the bands combination in color either using a color additive viewer or Diazo printer or by working off color processed prints of bands 4, 5, and 6 or bands 4, 5, and 7 at scales of 1:1,000,000 or 1:500,000 or occasionally 1:250,000.

If digital equipment is available, then the imagery analyst may prefer to work directly from the Landsat tapes after an initial examination of the overall area portrayed on black-and-white or color prints at scales of 1:1,000,000.

Geometrically corrected Landsat imagery will be of considerable use for matching the imagery to planimetric maps or in the production of semicontrolled mosaics at scales between 1:250,000 and 1:2,000,000. Possibly in the future, using Landsat-TM, SPOT, and ERS-1 imagery and the stereoscopic photographs of the metric camera on the Earth Shuttle, mapping will be possible at 1:100,000 or larger and map revision at 1:50,000. Map updating using Landsat MSS imagery is already undertaken experimentally at scales of 1:100,000 and occasionally at 1:50,000. Landsat MSS imagery, acquired at an altitude exceeding 30 times that of high altitude aerial photography, can be considered to be an orthographic projection of the earth's surface and hence may be used directly after geometric correction as equivalent to a small-scale (~ 1:250,000) planimetric or thematic map (Figure 8.3).

Vegetation recorded on Landsat color composites can usually be identified and classified into plant formations (see Section 4.4.2) based on differences in gray tone in band 5. This usually extends to the separation of grassland, woodland, shrubland, and forest, and the separation on two-seasons color imagery of the forest in the temperate northern hemisphere into hardwoods and conifers. A comparison of Landsat scenes taken in two different growing seasons is particularly helpful in studies of mixed land uses that include forest, bare land, and cereals.

Whether or not some plant subformations may be identified can only be determined by an initial study of the imagery associated with fieldwork. Success depends on unique characteristics of the subformations associated with major differences in their spectral reflectivity, their leaf-area indices, and their habitat. Possibly three forest or grassland classes may be identified. Although no breakdown of the plant formations may be achieved for tropical rainforest other than separating induced vegetation and regrowth forest, improved results may be obtained in low rainfall areas with considerable topographic relief. In general humid region landscapes have to be inferred mainly from vegetal variations visible on the imagery, which may only be occasionally cloud free. The best times are when vegetation shows maximum differentation between the photosynthesis of different plants in spring and autumn, or at the beginning of the rainy season, or at times of severe moisture stress such as during drought (Henderson-Sellers and Stockdale, 1979).

Landforms can be recognized on Landsat color composites mainly from structural, relief, and drainage patterns. These are clearest in arid areas and where the vegetation is relatively sparse. Seasonal differences are less important in the interpretation of landforms than in that of vegetation. Approximate heights and gradients in mountain areas can be inferred from the pattern of snow cover and from contrasts of light and shade on slopes exposed to, or oriented away from, the southeasterly solar illumination at the time of Landsat overpasses. Underlying rock structures are often observable at these small scales, usually allowing the discrimination of volcanic, plutonic, metamorphic, folded, faulted, and tilted uplands and their intervening aggradational lowlands. The pattern, direction, and density of drainage channels allow deductions to be made about valley gradients and the comparative roughness and composition of different materials, with soft or impermeable rocks having a denser network than hard or impermeable ones. It is sometimes possible to distinguish different lithologies. Figure 8.4, for instance, shows an area of tabular limestone in south Arabia whose dissection has led to the stepwise exposure of underlying beds of less permeable but softer rocks, and the formation of detrital valleys filled with light-colored alluvium, some of which is irrigated. Here the vegetation is mainly in the valleys. In other parts of the Middle East, such as the Zagros mountains of Iran, limestone hills carry a denser tree cover than do the neighboring shales.

8.5 SIDE-LOOKING AIRBORNE RADAR (SLAR)

The first successful SLAR Project of an extensive ground area (20,000 km^2) was the complete survey and mapping of the Darien Province (Panama) in 1968 using K-band radar. The province, because of continual cloud cover, had not been successfully photographed over a period exceeding 20 yr despite aircraft being available for most of that time. The SLAR imagery provided the first detailed planimetric maps of the province and valuable phytogeomorphic information, that had not been previously available.

Since then, projects in the tropics at scales of 1 : 125,000 or smaller have covered several extensive cloud-prone areas in Andean South America and Central America, Nigeria and Liberia in Africa, and parts of Indonesia and Papua-New Guinea in southeast Asia. By the end of 1980, about 8,500,000 km^2 of Brazil had been covered by SLAR, which includes the Radam project for the entire mapping of the Amazonian region. Studies and the small-scale mapping in Brazil have included vegetation and forest inventories, landform studies, and mineral exploration. In Nigeria the entire country was covered with SLAR imagery in the late 1970s. This has resulted in the preparation of national small-scale coastal zone land use maps showing drainage patterns and forest cover and considerable additional information on rainforest areas in the south of the country for which there was no Landsat imagery and which were only partly covered with very old aerial photographs at differing scales.

In general black-and-white SLAR has now proven itself to be economically important in providing small-scale planimetric and thematic maps containing phytogeomorphic information; and stereoscopic SLAR, comparisons of horizontally polarized and vertically polarized SLAR, and color recording to improve radiometric resolution have been used experimentally. Most maps are at a scale of 1 : 250,000 or smaller, although broad forest-type information and detailed land-use information can be derived from synthetic aperture imagery enlarged to 100,000 or possibly 1 : 50,000. Only occasionally are individual trees resolved and consequently there are few examples of tree species identification. An exception is the distinctive recognizable shape of mature *Araucaria cunninghami* trees along ridges on SLAR imagery of Papua-New Guinea.

As a generalization, maximum signals are returned from the ground to the aircraft by slopes facing the aircraft (compare Figures 8.5*a* and *b*), rough diffuse surfaces and surfaces, including dense vegetation, with a high moisture content; and minimum signals are returned from specular reflectors, such as calm water, and relatively smooth diffuse surfaces, including dry lake beds and wet surface soil. The X-band SLAR imagery, due to its partial penetration of vegetation, records many ground features including details of drainage patterns. Successful applications related to phytogeomorphology include mapping surface drainage patterns and wetlands, mapping major land use types, and the mapping of the surface expression of geologic features (e.g., folds, faults, joints, and other lineations).

Because SLAR usually has a much poorer resolution than aerial photographs, it has been used mostly for small-scale mapping and geomorphic study

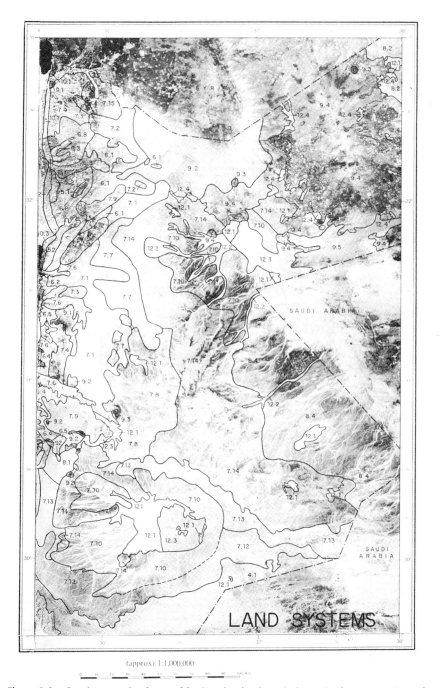

(approx) 1:1,000,000

0   10   20   30   40   50   60   70   80   90   100 Km

Figure 8.3. Landsat mosaic of part of Jordan showing boundaries and reference numbers of land systems interpreted on the imagery (after Mitchell and Howard, 1978).

Figure 8.4. Landsat view of the Jol Plateau area (People's Democratic Republic of Yemen). Note how the tabular limestone is eroded to reveal stepwise underlying softer but less permeable beds, which are darker in tone. The light-coloured detrital valleys have some irrigated tracts (NASA ERTS E-1168-06382 of 7 Jan, 1973).

(a)

(b)

Figure 8.5. SLAR imagery of part of the Berkshire Downs, England. Note the coarser resolution of the imagery when compared with the aerial photograph of a neighboring area in Figure 8.9. Note also the accentuation of features parallel to the line of flight: (*a*) north-looking, (*b*) south-looking. The line of escarpment across the image that is illuminated when viewed from the north but shadowed when viewed from the south is the north-facing escarpment of the Berkshire Downs near Wantage (Reading University photographs).

Figure 8.9. Single aerial photograph of part of the Berkshire Downs, England (cf. Figure 8.5). It shows land facets within two land systems (1) *Chalklands*: 12, moderate to steep slope; 13, dipslope plateau; 15, dry valley bottom; 17, chalk plateau; 20, moderate chalk slope. (2) *Clay vale*: 3, clay crest; 4, clay slope; 5, clay footslope; 8, small wet bottomland; 10, flood plain terrace. Land subfacets represent the subdivisions of such land facets that are sometimes visible, for example, side slopes and valley bottoms within land facet 8 and minor dry valleys within land facet 17. Note that it is possible to distinguish woodland (trees visible), arable fields (light tones), and pasture (dark tones). With stereoscopic cover it is also possible to distinguish conifers from hardwoods and some arable crops from others (U.K. Ordnance Survey photograph no. OS/66/85 of 30 August, 1966, Crown Copyright).

in parts of the humid tropics whose cloud cover virtually prohibits aerial photography. There are, however, certain types of surface detail which its low oblique angle of sensing reveals particularly well. It emphasizes small topographic contrasts in areas of low relief by highlighting the differences between those exposed toward and away from the sensor (Figure 8.5). This is especially valuable when an area is scanned from two different look directions. The SLAR also tends to emphasize square or linear features, especially when these are parallel to the line of flight. This is also illustrated in Figure 8.5. Despite its coarser resolution, the SLAR enhances such relief features as the escarpment of the Hampshire Downs running in an east–west direction near to the center of the imagery, which is illuminated on the south-facing scan but shaded on the north-facing scan.

8.6   AERIAL PHOTOGRAPHY

8.6.1   Diagnostics

The most important diagnostics for aerial photographic interpretation (API) of the ground objects recorded as images on aerial photographs follow (Howard, 1970a; American Society of Photogrammetry, 1960). In satellite imagery analysis, the identification and classification of objects depend primarily on the radiance values of the individual pixels:

1 Shape of the separate images (e.g., conifer crown).
2 Shadow characteristics.
3 Tone or color contrast (cf. radiance values in satellite imagery analysis).
4 Pattern provided by a group of the images.
5 Texture provided by the images.
6 Apparent height.
7 The natural site or habitat of the objects shown on the photographs.
8 Association of landforms and/or vegetation with other environmental factors.
9 Area of the object or group of subjects.

8.6.2   Shape

This relates to the configuration or the general outline of the object as recorded on the photograph. Some objects can be recognized on the single photograph, especially if the shadow outline can also be examined, but for smaller objects it is normally the three-dimensional or stereoscopic view that is so very important. Frequently objects have unfamiliar shapes when viewed vertically downward at the center of the photograph and the shape of the same object will change toward the edge of the photograph and also when viewed at greatly varying scales. Until analysts have familiarized themselves on single and stereo pairs of photographs with the characteristic shape of common objects at different viewing angles and different scales, they often find interpretation tedious and difficult.

Shape is important in recognizing ground features and woody vegetation. An example of the former is the characteristic shape of an oxbow lake found along some rivers (e.g., Mississippi). As regards vegetation two of the commonest shapes are the conelike crowns of most conifers and young hardwoods and the domelike crowns of many mature hardwoods.

8.6.3　Shadow

Normally fairly short shadows resulting from a sun angle over about 30° from the horizontal are preferred. When examining shadows, account must be taken of the time of the year, time of the day, latitude, and degree of slope of the ground. Shadow is helpful in identifying many small objects. In northern Canada and Scandinavia winter photography is often impracticable because of the long shadows thrown by the hills and trees.

For the study of open grown forests, deciduous forests when leafless, and some conifer forests, particularly with snow on the ground, shadow is frequently important. Narrow crowns, including the cone-shaped crowns of many conifers, do not resolve themselves fully on photographs, and measurement of the tree height may be more accurately determined from the shadow, particularly for open-grown forest with snow on the ground.

8.6.4　Tone and Color Contrast

Tone refers to the various shades of gray observed on black-and-white photographs, often used in delineating photographic communities–forest types at smaller scales. For example, in eastern Canada, *Sphagnum* sp. is characteristically a darker tone of gray than the tamarack trees growing in it and spruce is a darker tone again than the bog association. Color contrast refers to differences in hue, value, and chroma and provides up to 200,000 observable differences (Becking, 1959). Closely associated with tone is the tonal contrast between adjoining images and the edge gradient of the images.

Consensus of opinion varies concerning the faculty of perceiving tonal differences. Kodak markets a tonal scale with 10% differences in tone. Losee

(1951) used two observers to see what monochromatic tonal differences they were able to distinguish within an aperture of 5 mm diameter. He concluded that to obtain 95% success in identifying tonal differences the experienced observer requires a difference in gray-scale density of 0.21.

Location of an image on a photograph relative to its principal point can also influence tone. Trees of the same species and some landforms may appear progressively darker in tone the farther they are from the plumb (nadir) principal point. It is also known from optics that a small gray object appears much darker on a white background than on a dark background.

Color contrast in color photography is equivalent to tone in black-and-white photography. A slight change in color, as expressed by its hue, value, and chroma, is usually more readily recognized on color transparencies than the difference in tone on black-and-white photographs. Color photography is helpful in species identification and soil type and landform recognition where the vegetal cover is low. Color contrast in the infrared is important in separating dead and living matter and in separating moist and dry landforms.

8.6.5 Texture

Texture may be considered to be a microcharacteristic corresponding to pattern as a macrocharacteristic. Texture can be described as the frequency of tonal change within the images on the photograph. It is produced by an aggregate of unit features too small to be clearly discerned individually. It is the product of tone, size, shape, pattern, shadow, and reflective qualities of the object and varies with the photographic scale. As the scale is reduced, so does the texture become finer; and what may be considered as pattern on larger scale photographs provides texture on a very small-scale photograph and on satellite imagery texture is no longer observable. On very large-scale photographs, for example, 1:2000, individual groups of leaves contribute to texture. At 1:5000 leaves and branches contribute to the texture; at 1:10,000 tree crowns provide the texture, at 1:30,000 it is the individual trees and groups of shrubs; and at 1:80,000 small land facets, land elements, surface lineations, and entire plantations that contribute to texture. On black-and-white photographs a 50:50 division of black and white into extremely small units provides a fine texture. The larger and more irregular the units, the coarser the texture will be. Density scales are closely associated with texture. They are commonly used in plant ecology studies to estimate the crown cover of trees per unit area (Figure 7.1).

8.6.6 Pattern

Pattern is the broader arrangement of tones and texture and is associated with local landforms, geology, topography, drainage networks, plant formations, and associations and human activity. Pattern is a macrocharacteristic used to describe the general spatial arrangement of the images. Particularly in studying landforms, it is important to understand and to recognize patterns produced by morphogenic processes. In tropical areas, a difference in the vegetative pattern often records the past presence of a human population long after it has abandoned the area. Pattern often shows up well on mosaics, being strongly influenced by the landforms. Poor drainage, for example, on basalt provides a distinctive pattern. Common drainage patterns are illustrated in Figure 8.6.

8.6.7 Apparent Height

One may ask why the adjective apparent is used. The explanation is that there are a number of factors influencing the determination of the height of objects from the photographs that result in the measured or calculated photographic height being different from the actual height as measured on the ground. Provided these sources of error are recognized, then the true height, or a reasonable approximation of the true height, can usually be determined from the apparent height by adding a correction factor. Apparent heights are calculated

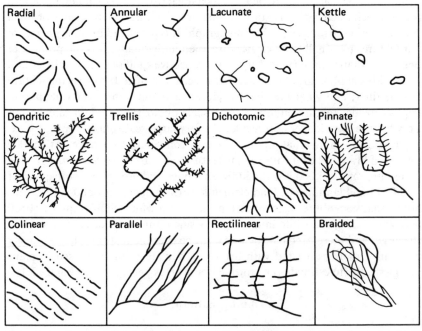

Figure 8.6. Common drainage patterns observed on aerial photographs and associated with landforms at scales of ~ 1:10,000–1:100,000. These can be given descriptive names as shown.

from the measurement usually made on stereoscopic pairs of aerial photographs, and the correction factors are determined for systematic errors by comparing the photographic measurements with field measurements. The height of relatively large objects recorded as images on single photographs may be determined roughly by radial displacement and more accurately by measurement of shadow length. Parallax difference measurements from stereoscopic pairs of aerial photographs are used to determine the apparent heights of individual trees and forest stands, point elevations, the photographic scale of small areas, slopes, and the dip and strike of exposed geologic structures.

It is important to appreciate that the height of an object is exaggerated in the steroscopic image. This is both disadvantageous and useful. To the unwary it may lead to an incorrect visual assessment. To others it is a more sensitive means for certain purposes of comparing differences in height. For example, topographic differences in height are more conspicuous from photographs than from a vantage point on the ground or from an aircraft. This may be helpful when studying the influence of site factors. Stand height above a shrub layer and isolated trees and coppice above a herb layer are advantageously exaggerated, and dominant trees are often conspicuously taller than codominant trees of the main canopy layer. Under the stereoscope a slope may appear to be two or three times greater than it is on the ground.

Slope is of interest when determining the true size of microareas, site quality in forestry development, the strike and dip of geological structures, sometimes the boundaries or location of plant communities, and in planning access to rural areas. In rugged terrain, slope may introduce a considerable error in relation to microareas, including sample plots. The slope observable on a stereoscopic pair of photographs can be determined using either a parallax bar, a parallax wedge, or a slope wedge. Using a parallax bar or wedge, two spot heights are separately calculated. One is at the top of the slope and one at the bottom of the slope, as observed in the stereoscopic model. Their difference or vertical interval provides the difference in vertical height V. The horizontal distance H between the points is also determined either by direct planimetric or topographic map measurements or by photographic measurements using radial line intersections. Thus slope percentage S is provided by

$$S = \frac{H}{V} \times 100$$

Apparent height differences as observed in the stereoscopic models and as obtained by using certain types of photogrammetric plotting instruments can provide stereoscopic profiles across the landscape. The basic technique was used as long ago as 1939 by Hugershoff in Germany to provide profiles of forest stands (Spurr, 1960). Figures 8.7 and 8.8 illustrate the method applied to pro-

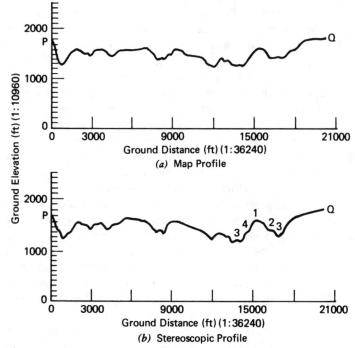

Figure 8.7. Landform profiles. To supplement thematic maps, scale profiles across the terrain are useful. These may be prepared (*a*) directly from a topographic map or (*b*) with the aid of a suitable photogrammetric instrument from stereoscopic pairs of aerial photographs. Note the increased detail provided by the use of aerial photographs and that land facets can be identified (points 1–4) (Howard, 1970b).

viding stereoscopic profiles of land units and woody vegetation (Howard, 1970b,c). These profiles complement block diagrams used in geomorphology and vegetation transects used in vegetation studies.

8.6.8   Site

The term site has two distinct meanings. First, it is widely used in the study of aerial photographs to describe the location of the image under observation in relation to its surrounding features. Its more restricted meaning connotes the sum of the factors of the environment influencing plant growth. In eastern Canada, for example, black spruce can often be identified on aerial photographs by knowing that the single species is associated with low quality *Sphagnum* bog, which is readily identified on the photographs by its site, shape, and tone or color.

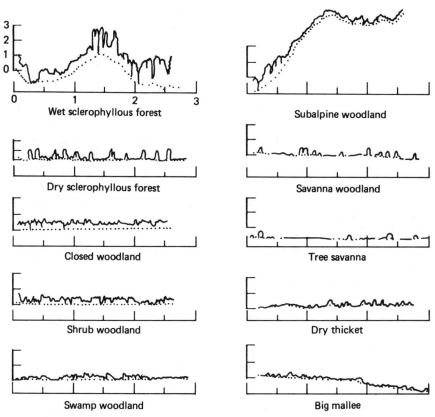

Figure 8.8. Vegetation profiles obtained from stereoscopic pairs of aerial photographs of a major plant subformation in Victoria, Australia. The *x* axis provides the ground distance in thousands of meters, whereas the *y* axis indicates heights in 30-m steps above the ground datum (dotted line) (Howard, 1970c).

Aspect, topography, landforms, soil, and the characteristic natural vegetation are all important environmental factors when examining site. The first three may be classed as macrocharacteristics and the last two, with soil moisture and tree size at maturity, may be considered as microcharacteristics. The relative importance of each varies with local conditions. Thus at high latitudes aspect is often the most important. In tropical America, elevation and rainfall probably take precedence. In maritime United States, geology, aspect, and elevation are important. In photo interpretation, aspect, slope, landform, and the natural vegetation provide the key factors in the preparation of site maps from photographs.

8.6.9   Association

This is another term with two distinct meanings. In its phytogeomorphic context it refers to a plant community of definite floristic composition that presents a uniform physiognomy and grows in uniform habitat conditions. In general aerial photographic interpretation the term has a much wider meaning. In continuing from the examples and discussion on site, it will be appreciated that some objects are so closely associated with others that each helps to confirm the presence of the other. The term *correlations* is sometimes used instead of association.

Only one class of object may be discernible on the photograph; but its shape, tone, pattern, texture, area, height, and/or site are associated with another class of object not recorded or not clearly recorded. By studying one or more of these collateral characteristics that have been observed to be associated with the object not clearly recorded, characteristics of the latter can be evaluated. Thus the forest type may be so closely associated with the landform and soil type that the soil type can be delineated by the boundaries of the forest type as seen in stereovision.

As early as 1928 Zieger observed in studying aerial photographs in central Germany that *Pinus sylvestris* (Scotch pine) with shorter boles and wider crowns was associated with poor sites. In Victoria, Australia, *Eucalyptus camaldulensis* of higher site quality is associated with the flood plain of the Murray. This association can be inverted for use by hydrologists to give them a useful tool; that is, where the higher quality red gum occurs near the Murray one is in an area liable to flooding either annually or every two or three years. Some landforms guide in the recognition of soil textures. Fossil channels and abandoned river meanders, for instance, although dry and often imperceptible on the ground, can be recognized on aerial photographs. Arborescent networks visible from the air in the English fenlands have been identified as old river courses exposed by the subsidence of the peat, and have sandier soil textures than their surroundings (Seale, 1975).

8.7   THEMATIC MAPPING USING AERIAL PHOTOGRAPHS

Frequently horizontal linear and areal measurements, corrected for scale, are taken directly off a single photograph. These measurements are very approximate because the photograph is a perspective view and not an orthographic projection; thus they show some radial displacement and are liable to additional errors toward the edges of photographs and in rugged terrain. Again, if the aircraft has tilted excessively at the time the photographs were taken, radial displacement may be considerable and has to be corrected during photographic processing or mapping.

An assemblage of the aerial photographs as an uncontrolled photographic mosaic contains the errors inherent in each photograph and the accumulation of the errors between photographs. These errors are reduced in a semicontrolled mosaic by adjusting the photographs to fit an overall accurate map base and by using only the center area of each photograph when preparing the mosaic.

In general, using aerial photographic mosaics and stereoscopic pairs of aerial photographs and single photographs, landforms can be identified, their boundaries mapped, and the natural vegetation divided into plant formations and plant subformations, and the landscape divided into land units (Figure 8.9). The success in identifying and mapping the plant subformations will depend considerably on the resolution of the aerial photographs. Usually, with older aerial photographs at scales of 1:25,000 or larger and at smaller scales with modern color infrared photographs, it is possible to examine the structure of the woody vegetation and to classify a range of subformations. Several height classes of woody vegetation can be assessed stereoscopically, and depending on the scale and quality of the aerial photographs their stand heights measured stereoscopically in 2 to 7 m classes, their crown cover estimated in 10-20% density classes, and crown widths measured with a precision of 2-5 m. From these parameters and other photographic parameters much of the natural vegetation within major land units can be classified in terms of its physiognomy portrayed on aerial photographs (Table 8.2). At scales larger than about 1:15,000 the height of the dominant woody vegetation can be stereoscopically measured. Species identification usually improves when color photography is used at scales larger than about 1:8000 to 1:10,000. In eastern Canada many tree species have been identified on very large-scale photographs, for example, 1:2000 (Sayn-Wittgenstein, 1961).

In a comparative study of aerial photographic film–filter combinations and scales applied to land unit classification and their mapping in central Victoria (Howard, 1970d) the following results were obtained. Surprisingly most of the land facets could be recognized in the smallest scale aerial photographs (1:84,000), although the stand parameters of the woody vegetation could not be accurately measured at scales smaller than 1:15,840 and preferably at 1:10,000 or larger. In contrast, land systems were more readily identified at scales between 1:30,000 and 1:84,000, but the identification of land subprovinces and land provinces needed aerial photographic mosaics (or now Landsat imagery). When film–filter combinations were examined in relation to the landform or vegetation of the land units, their impact was not as great as might be expected. Black-and-white infrared photography was found the least useful due to the loss of detail in shade. Color infrared has a somewhat similar disadvantage, but was much more useful because of color variations associated with changes in vegetal structure, leaf area index, and canopy cover. In the hands of a skilled photo interpreter panchromatic black-and-white proved it-

Table 8.2 Principal Subformations and Their Photogrammetric Parameters in Southeastern Australia.

Formation	Subformation	Stand Height (meters)		Crown Closure (%)		MS Crown Width	H:MSCW Ratio	Photographic Tonal Range	
		Main Story	Under-Story	Total	Main Story			MS	US
Forest									
Rainforest	Temperate	30 ± 8	11 ± 6	~100	35 ± 10	13 ± 3	2.5	0.2-0.7	>0.7
Sclerophyll	(Wet scler)	55 ± 16	16 ± 10	90 ± 10	75 ± 15	15 + 5	>3.2	0.3-0.1	>0.5
Forest	(Dry scler)	35 ± 8	<5	65 ± 20	60 ± 25	13 ± 5	2.4	0.3-0.0	<0.5
Woodland	Dense (closed)	18 ± 6		45 ± 15	45 ± 15	10 ± 3	1.8	0.3-0.5	<0.3
	Layered	18 ± 5	10 ± 2	50 ± 10	30 ± 10	11 ± 2	1.6	0.3-0.7	>0.3
	Shrub	16 ± 3	<5	50 ± 10	35 ± 10	11 ± 3	1.5	0.2-0.5	<0.4
	Swamp	16 ± 5	<5	35 ± 10	35 ± 10	11 ± 3	1.6	0.3-0.7	>0.3
	Open (savanna)	15 ± 3		25 ± 5	25 ± 5	13 ± 2	1.1	0.5-0.7	<0.2
	Heath	12 ± 3	<2	35 ± 10	35 ± 10	10 ± 2	1.2	0.5-1.0	>0.5
	Tree savanna	10 ± 3		30 ± 15	30 ± 15	8 ± 2	1.2	0.3-1.0	<0.5
	Subalpine	10 ± 2		15 ± 10	15 ± 10	8 ± 2	1.2	0.5-0.7	<0.3
Tall Shrubland	Thicket (wet)	8 ± 3		60 ± 15	60 ± 15			0.7-0.1	>0.7
(scrub)	Thicket (dry)	6 ± 3		45 ± 15	45 + 15			0.6-0.8	<0.7
	Big mallee	8 ± 2	<2	50 ± 10	50 ± 10	<5	<1	0.7-1.0	<0.3
	Mallee	6 ± 2		30 ± 10	30 ± 10	6 ± 2	<1	0.5-0.7	<0.1
	Savanna	6 ± 2		25 ± 5	25 ± 5	6 ± 2	<1		<0.2
	Mallee heath					5			
Low Shrubland	Shrub-steppe	<2		<40	<40	<5		<0.3	<0.1
	Heath (wet)							0.7-0.3	
	Heath (dry)							0.3-0.7	
Herbland/ Grassland	Several grassland types							<0.2	

^aFrom Howard (1970c). MS = main story; MSCW = main story crown width; US = understory.

self remarkably useful for studies of the smaller land units excluding land elements. Overall, color film was found to provide most data related to phytogeomorphology and panchromatic color was preferred.

On photographs at a scale of 1:84,000, it was possible to identify the boundaries of part of the land regions, land systems, land catenas, and some land facets. In addition, color photographs at 1:84,000 indicated that land units are more readily interpreted in color; but an experienced interpreter using a zoom stereoscope was often equally efficient with black-and-white photographs. In areas of sparse vegetation exposed soil types could often only be delineated on color photographs. At large scale the increasing number of photographs provide a practical problem; and the boundaries of the areas defined by distinctive drainage networks become increasingly inconspicuous. On color photographs the differences in hue, value, and chroma made interpretation easier, but no more definitive drainage characteristics could be extracted than from panchromatic black-and-white photographs, although sometimes major drainage lines are more readily observed on infrared color prints. As reported elsewhere (Howard, 1970b), quantitative expressions of the drainage network (Strahler, 1964) can be extracted from the aerial photographs (e.g., drainage density, bifurcation ratio, hypsometric frequency). Also, within land systems land catenas were identified and represented by the stereoscopic profiles (Figure 8.7; Howard, 1970b). On the multiband photographs, the land catenas were more readily identified on color than on black-and-white prints.

When the land facets were examined on the multiband photographs, it was observed that the land facets and their boundaries are sometimes best observed on infrared color photographs, but unfortunately the effect of increased haze, major changes in flying height, and minor changes in processing controls could considerably alter the color balance. This gives a certain unpredictability to the photo interpretation of infrared color prints that does not occur with panchromatic color and panchromatic black and white.

NINE

FIELD SURVEY

9.1 PREPARATORY ACTIVITIES

9.1.1 Introduction

Phytogeomorphologic surveys should begin with a plan that incorporates the overall objectives and an estimate of the cost and main steps by which they are to be achieved. Normally this includes the specification of the boundaries of the area to be covered, the level of examination (scale), the statistical methods and sampling design, the description of the fieldwork and of the laboratory analyses, and the identification of subsidiary investigations.

The next stage is one of organization and finalization of the detailed budget. In large surveys this will include identifying the type of experts required, the recruitment of staff, and decisions on the program to be followed to achieve the main objectives. Planning is aided and more easily envisaged by preparing *flow diagrams*, *bar charts*, and *critical path analyses*. Flow diagrams are simple graphic illustrations that show the sequence and interactions of different activities in a project in a systematic way. An example of such a flow diagram illustrating the stages of a hierarchical subdivision of landscape is shown in Figure 6.2.

Bar charts are primarily timetables showing the phased inputs of various specialists in a program and are used specifically to analyze and program the staffing, as shown in Figure 9.1. Bar charts are valuable in demonstrating

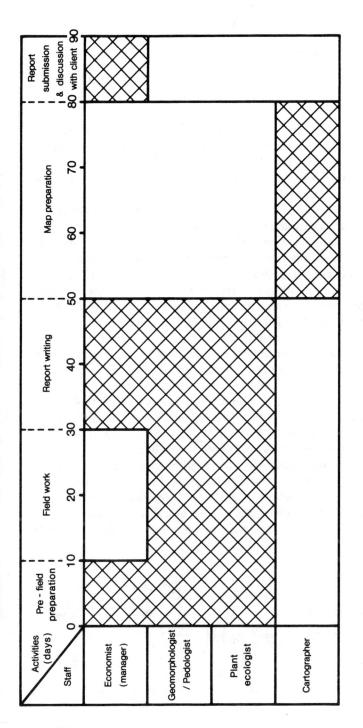

Time free

Time engaged

Figure 9.1. Bar chart showing a program for 4 specialists to carry out a phytogeomorphic survey for land development in 90 days.

clearly the times when specialized staff are required and help in coordinating their activities, moves, and contacts.

9.1.2   Critical Path Analysis

This is a further development to assist in the planning of project activities by dividing the major activities into their components and depicting them in the form of a network showing their detailed timewise relationships through the project. The network can be used for optimizing the utilization of resources, progress reporting, and control.

The procedure, as outlined by Dale and Michelon (1966) and Baboulene (1969), is to construct a diagram in which events, represented by circles, are linked by activities represented by arrows (see Figure 9.2). Each event marks the completion of one or more activities and must occur before other dependent activities can begin. Durations can be given for activities and dates for events that, if they are imposed, can be flagged. As shown in Figure 9.2, planning cannot be finalized until *both* staff and equipment are available. Event (1) is thus a *merge point*. On the other hand, neither the collation of field observations nor the identification of plant samples and the analysis of soil samples can normally begin until the workers return from the field survey, and so event (2) is a *burst point*. This network could be considered as consisting of a number of subnetworks, one of which might be the subdivision of activity ② into its component operations such as the obtaining of transport, aerial photographs and satellite imagery, maps, and survey instruments. Where an event is also a decision, it is separated as a *decision box*, as shown on Figure 9.3. If new aerial photography is required, then this would precede the burst point.

The *critical path* is the longest time sequence through the network, because this is the path that controls the time of final completion. Events and activities along this are known as *critical events* and *critical activities*. The critical path is marked by doubling the activity lines along it as shown in Figures 9.2, 9.3, and 9.4. Where the diagram is large it is often necessary to redraw it to make

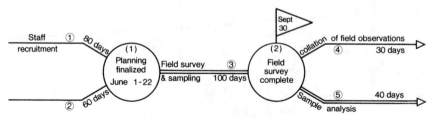

Figure 9.2.   Example of a network of working operations for a survey illustrating the use of critical path analysis.

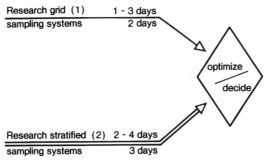

Figure 9.3. Section of a critical path analysis illustrating a decision box and maximum, minimum, and average time indications.

this line central. The *float* is the time available for an activity in excess of its necessary duration, which is, for example, 20 days for activity ②. It is often desirable to put in the most optimistic and the most pessimistic estimates, and the most likely times, rather than show only a single time for each activity, as shown in Figure 9.3. The *slack* is the difference between the earliest and the latest times for an event, both of which can be shown within the event circles. In Figure 9.2 the slack at event (1) would depend on the starting times of activities ① and ②. Figure 9.4 illustrates a rather more complicated critical path analysis for producing a survey report. In complex forms of critical path analysis, computers can be used to generate and update diagrams and to make calculation of staff requirements and the timing of resource acquisitions.

Use of the method has a number of advantages. It ensures that planning is comprehensive and greatly reduces the risk of overlooking events. It permits calculation of whether individual jobs can be completed in time and focuses attention on potential bottlenecks and where the application of resources can be most effective. It shows which activities can be carried out simultaneously and where there is likely to be spare time. Finally, it makes it easier to monitor progress, especially in detecting slippage along the critical path.

9.2 SAMPLING STRATEGY

9.2.1 Introduction

Before fieldwork is undertaken, it is necessary to exploit fully all available information from published and unpublished sources. This necessitates the allocation of a time period for the study of archives and the preliminary analysis of remotely sensed imagery, usually the most up-to-date source of synoptic information against which data can be checked.

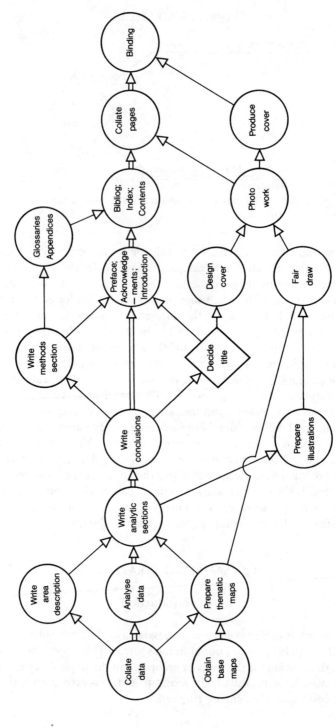

Figure 9.4. Example of critical path analysis applied to the final stage of a phytogeomorphic survey, from completion of field work to final report production.

The strategy for sampling must avoid bias and allow for replication to ensure rigor in the statistical analyses. Both the sample design and the details to be recorded are governed mainly by the purpose of the survey, the resources available, the desired accuracy and precision, and the quality of the remotely sensed imagery that can be used for the extrapolation of field data (see also Section 8.2).

The number of field sites and samples and the sampling design vary with the objectives and scales of the survey and the required accuracy and precision. For landform reconnaissance at mapping scales smaller than about 1:1 million, analysts with local experience can often achieve their geomorphic objectives by relying on archival information in conjunction with remotely sensed imagery. Otherwise the sampling design, particularly as related to vegetation, must include collection of data in the field.

9.2.2 Stratification

The purpose of stratification is to reduce the variation within each stratum and to increase the variation between strata by dividing the landscape, as observed on maps, mosaics, and stereo pairs of photographs, into more homogeneous classes. Alternatively, stratification may be viewed as giving a reliability similar to unstratified sampling but with fewer samples. The value of stratification cannot be overemphasised. For example, in parts of Thailand, Loetsch (1957) found that the results obtained using stratification would have needed four times as many plots without stratification to have given comparable results. If the area is not covered by remotely sensed imagery, it is usually not possible to take advantage of the increased efficiency that is obtained by stratification prior to sampling; but statistical analysis after random sampling may suggest that the data should be grouped into strata prior to calculating standard errors and confidence limits. Strata based on land systems usually provide a sound basis for vegetal studies.

Assuming the same number of samples are used to test the relative efficiency of stratifying and not stratifying an area, a useful indicator is provided by the variance ratio or F-test (see Husch, 1963) which is obtained by dividing the between-strata variance by the within-stratum variance. The variance S^2 is obtained by squaring the standard deviation S. If the several strata represent a single population, the two variances will be similar. Other techniques using variance have also been suggested. For example, Moessner (1963) has suggested an approximate method of rating a number of stratification schemes and providing an estimate of the number of samples needed to give similar reliability using different methods of stratification. The pooled within-stratum variance of each scheme ($\Sigma P_i S_i^2$), obtained by weighting the variance of each stratum within the scheme by its area (P_i expressed as a percentage of total area), is divided by the variance of the unstratified scheme S^2. This is then subtracted from one, that is,

$$1.00 - \Sigma P_i S_i^2 / S^2$$

The remainder is then used as rating of the efficiency of the method of stratification and as an estimate of the reduction in the number of plots needed to give the same reliability as an unstratified scheme.

Avery (1964) gave three methods for determining the number of samples required in a forest survey by calculating the variance of each stratum or type. The methods assume aerial photographs are used and that strata can be recognized. The first method, the most obvious, involves laying out a series of random samples from photographic strata in the field and using these to calculate the variance of each stratum prior to a full survey. This, however, is costly and often impracticable due to the terrain conditions. A modification of this method involves allocating field plots directly to photographic strata according to the sampling precision of similar surveys elsewhere; but there are statistical objections to this approach. The second method involves the full computation of the variances from photo plots; but often can only be used if large-scale photographs and other relevant data are available. For example, in forest surveys of timber volume, not only are photographs needed at a scale larger than about 1/16,000 but areal volume tables are also necessary.

A third method calls for the delineation of strata on the photographs, and then by stereoscopic examination an estimation is made of the range of values in each stratum. The range in each stratum is divided by four to indicate what the standard deviation is likely to be; and this is squared to provide a crude estimation of the variance. The method assumes the samples are normally distributed; and as pointed out by Avery, the reason for dividing by four is readily appreciated if one recalls that a dispersion of 95% of the samples encompasses ± 1 standard deviations.

The crude estimations of the standard deviation and variance so obtained are then used to calculate how many plots should be allocated to each stratum. For sampling in the field to give approximately the same standard errors of the mean, the calculated variances are summed (ΣS_i^2) and the variance of each stratum is expressed as a percentage of the total

$$(S_i^2 \times 100)/\Sigma S_i^2$$

For a predetermined number (n) of plots, the number n_i to be allocated to the ith stratum to provide a similar standard error of the mean will be

$$(S_i^2 \times 100 \times n)/\Sigma S_i^2$$

The standard error of the mean will be

$$\sqrt{\text{variance}/n_i}$$

In practice this method may result in too few plots being allocated to a stratum estimated as having the greatest range. Also, it may be desired to place varying confidence limits on the mean of each stratum according to other criteria. For example, in forest surveys each stratum based on tree species and stand volume in cubic meters or cubic feet may have a different market value; and as a result, it is desirable to vary the standard error of the mean with value of the forest stand as ascertained by stereoscopic examinations of the photographs. Under these circumstances, the crude estimate of the standard deviation S_i is used to calculate the number of plots to be allocated to each stratum for a specified sampling error. Sampling error is the standard error $S_{\bar{x}}$ expressed as a percentage of the mean $\bar{x}$, that is,

$$(S_{\bar{x}}/\bar{x}) \times 100$$

In forest surveys, sampling errors of 5, 10, 15, and 20% are commonly adopted.

The estimated mean of the stratum multiplied by the sampling error will give a crude estimate of the standard error of the mean ($S_{\bar{x}} = \bar{X}$) (sampling error), and hence an estimate can be made of the number of plots n to be allocated to each stratum.

For the ith stratum

$$n_i = \frac{(S_i)^2}{(S_{\bar{x}_i})^2}$$

at the 68% probability level, or

$$n = \frac{(S)^2}{(S_{\bar{x}})^2}$$

for an unstratified population, or

$$n = \frac{(S)^2}{(S_{\bar{x}})^2} \frac{N - n}{N}$$

for an unstratified small finite population where $N =$ maximum number of samples in the population.

If the probability level is increased to 95%, 3.84 times the number of plots will be required, because t^2 for a large number of samples is 1.96^2 ($= 3.84$).

The method of weighting the allocation of samples by area has been commonly used in forest inventories; but has the disadvantage of providing standard errors of the mean that may be out of keeping with the objects of the survey. Obviously strata weighted by area and covering large areas will be allocated more sample plots than strata covering smaller areas; and as a cursory examination of a t-table will show, the t values used in fixing the confidence limits of the standard error of the means vary according to the number of samples.

A factor often overlooked is that efficiency, using standard error as the criterion, will not be greatly increased by using more than 40 to 60 plots unless the number of plots is greatly increased. Possibly the error will be halved by quadrupling the number of sample plots. The hypothesis may be formulated that within any one stratum the number of plots should normally not exceed 40 to 60. The word normally is used because, if the population is suspected of being heterogeneous, the number of plots must be increased beyond these limits in expectation that upon being statistically analyzed the data will be broken down into two or more strata. For example, if a stratum is suspected of being heterogeneous and likely to provide three homogeneous strata, then up to 120 plots (3 × 40) might be allocated initially.

9.2.3 Sampling

There are four major types of sampling: purposive, systematic, probability or random, and cluster.

Purposive sampling seeks to identify and to collect data within typical sample sites that are considered representative of distinct landscape units, and are identified from remotely sensed imagery. It makes use of background knowledge and local expertise to stratify an extensive ground area into what are judged to be relatively homogeneous units. Thus it takes into account all the important variations in the landscape within optimally accessible lines or strips of plots; however, it has the major disadvantages that it relies on a subjective selection, the tally of sites observed gives no automatic indication of their representativeness of the populations being sampled, and statistical confidence limits cannot be established for the calculated means of the populations.

Systematic sampling is a modification of purposive sampling that uses samples regularly spaced at grid intersections or that are regularly spaced along line or strip transects. It is thus objective in selection, but otherwise has the weakness of purposive sampling. However, random samples can be introduced by giving a random start to each grid or group of transects. This gives coverwise an even distribution of samples, but it may introduce bias if both the terrain and the grid are congruently directional. In addition, the number of degrees of freedom is limited to the number of random starts and the not the total number of "sample plots." It is basically a form of cluster sampling with few degrees of freedom. In practice the difficulty of ground access may make it hard to achieve randomness of samples. Where this occurs it is best to retain the principle that the distribution of samples at least avoids bias from the na- ture of the terrain itself.

At present the consensus of opinion suggests that systematic sampling will provide as good or sometimes even a better estimate of the mean for a specified number of samples than random sampling. However, as probability theory is based on random selection, the theory and techniques of random sampling no longer apply; and the calculation of standard error of the mean, variance ratio, fiducial limits, and other statistical parameters are no longer valid and there- fore remain unknown. In practice, as it is often expensive or time consuming to locate random plots in the field, purposive or systematic sampling may be preferred. According to Seely (1964) an estimate of standard error of the mean can be made from systematic samples that are only moderately biased. The final decision, however, on whether to use random or systematic methods must depend on local conditions, on the way in which the collected data is to be analyzed and used, and on the objectives of the survey. Rosayro (1959) com- pared random sampling and samples taken along lines drawn between promi- nent map features (selected line sampling) and found there were no important differences.

Probability or *random sampling* on the other hand selects study areas and sites from the population in such a manner that each has a known chance of appearing in the sample. Where the chances are made equal, this leads to sim- ple random sampling in which sites or areas are chosen on a numerical array covering the area on the basis of random numbers. However, this has the dis- advantage that it takes no account of the variations in the landscape or of the value of concentrating efforts into important or complex areas and may result in omitting samples of areas of importance. This can be overcome however, by combining random samples (preferably in pairs) with stratification.

The fourth major type of sampling is *cluster sampling*. If a group of plots provides a fixed pattern in relation to each other, and the group as a unit pro- vides part of another pattern formed by other units, then each group of plots is referred to as a *cluster*. Frequently plots in a cluster are systematically arranged and the clusters as units are randomly arranged. Plots in a cluster are often ar- ranged in a line or on the sides of a square, rectangle, or triangle or in the form of a cross. In Sweden for the third national forest survey, the clusters of plots were arranged on a square (Hagberg, 1956). In Thailand Loetsch (1957), using photographs at 1/48,000 employed clusters of seven systematically arranged squares (camp units) as a compromise between maximum dispersion of plots and forest conditions. One random photopoint only was needed, that of the the center square. Forty-eight $\frac{1}{2}$ hectare sampling areas were located around the periphery of each square.

Provided each cluster of plots is randomized in relation to the other clusters, statistical parameters can be calculated even though the plots within a cluster are systematically arranged. The number of degrees of freedom, however, will be one less than the number of clusters ($m - 1$) and not one less than the number of plots. Assuming there are m clusters and n plots in each cluster, then the variance S^2 is

$$S^2 = \Sigma \frac{(\bar{X}_j - \bar{X})^2}{m - 1}$$

where X_j = mean of the jth cluster and $\bar{X}$ = mean value for all plots, that is,

$$X = \frac{X}{mn}$$

The standard error of the mean for all plots in all clusters will be

$$S^2 = \sqrt{\frac{S_2}{m} \times \frac{M - m}{M}}$$

where M equals the maximum number of clusters in the population and corresponds to N samples in a finite population. If this is compared with the standard error, assuming all plots are randomized, then

$$S^2 = \sqrt{\frac{S_2}{n} - \frac{N - n}{N}}$$

Observe that degrees of freedom are lost and therefore statistically the method is not as efficient.

When the sample design is in the form of many equally distanced plots along lines or strips and each line or strip is randomized, then each strip–line transect provides a cluster sample and the degree of freedom is one less than the number of transects and not one less than the number of plots.

9.2.4 Size of Sample Plots

It is advantageous to consider carefully the size of sampling units to be used because this has an important bearing on statistical results.

The selection of sampling sizes suitable for the study of geomorphic phenomena depends on the scale being considered, the complexity of the landscape, and the nature of the important surface processes. In general scale can be related to the appropriate level in the hierarchy of geomorphic units so that, for instance, at 1:250,000 one would select representative samples of land systems, at 1:50,000 select representative samples of land facets, and so on. A larger number of samples will be required at any scale to represent the variance where the landscape is complex. Where a process such as fluvial or wind action is important, the sample should be so arranged as to include the whole area in which the process operates or a logically selected part of this area. For instance, to sample fluvial erosion one must study a catchment or subcatchment, or to sample wind erosion one must sample the area covered by the particular wind regime, or some recognizable part of this caused by, for instance, the arrangement of certain major obstacles.

When selecting the size of sample areas suitable to the study of the distribution of vegetation types, it is important to remember that there are often significant interrelations in small areas between plot size and the pattern provided by the under- and overdispersion of plant species, (see Greig-Smith, 1964). For example, in the Brazilian tropical rainforest, according to Heinsdijk (1960), tree species seem generally to be randomly distributed, but group sizes of 2–5 ha were detected.

In forest surveys, using aerial photographs, the sizes of sample plots are normally $\frac{1}{25}$, $\frac{1}{20}$, $\frac{1}{10}$ and $\frac{1}{5}$ ha, although smaller plots have been used for very young stands and $\frac{1}{2}$ ha plots for large trees in old–overmature forest. In herbaceous plant community studies, quadrat size is commonly 1 m^2 or subdivisions or multiples of this unit, or alternatively, species may be recorded along a line transect or strip transect.

It is sometimes desirable to vary the size of the sample plots according to species or vegetal types in an attempt to provide normal distribution curves. This may be termed *variable plot sampling*. For example, during the collection of field data in Wales relating to the diameter classes of Japanese larch, a fixed plot size resulted in a large number of measurements for small and medium-sized trees but few for the sawlog-sized trees in the same compartments. By increasing the plot diameters for the larger trees, zero readings at many of the plots could be avoided and thus an adequate number of samples obtained for statistical analysis. In Tanzania, Howard (1959) used, in conjunction with strata based on land systems for forest inventory, a plot design of four concentric circles (0.1–0.25 acre). Sampling intensity varied between 0.1 and 1.2%; and the size of the plots used for each tree species depended on its estimated frequency and market value–market potential. Plots were located in clusters on line transects and these were randomized in pairs within blocks of 15 to 30 square miles. If the separate sampling strata of land units or individual samples (single sample plots, cluster samples, or strip transects) are to be plotted on a map, then the final mapping scale needs to be taken into consideration and 2 mm is about the narrowest strip and 3 × 3 mm the smallest area that it is practicable to separate in mapping. At a scale of 1 : 50,000 these represent widths of 100 m and areas of 150 × 150 m, as being the minimum areas, that is, the mapping cells that the ground sampling should represent.

As we progress with the design and improvement of the sampling method, the impact of the distribution pattern of the various plant species will need to be considered as well as the size and number of the samples. Sample design may include the collection of associated environmental data (e.g., slope, aspect, soil type), the weighting of recorded species by their basal area, leaf cover, and so on, and the distance between nearest individuals (e.g., nearest neighbor distances).

9.2.5   Choice of Parameters in Statistical Analysis

The range of techniques is wide and it is necessary to select those most appropriate to the particular investigation. They are important at two stages: first, to ensure that the sampling scheme is likely to provide the required information and second to analyze the data when it is obtained.

For most purposes individual values in a group are considered to have a *Gaussian* or *normal* distribution about their mean. When this is not so, their distribution is considered to be *skewed*. Because of the number of techniques that require a normal distribution, it is desirable, whenever possible, to transform the individual items of data to bring them to this form.

In the analytical stage statistics serve three major purposes. First, they provide a numerical description of the population of sites and their phytogeomorphic characteristics. The *arithmetic mean* indicates the average of a group of values. For some purposes the *median*, the middle value of those observed, or the *mode*, the most frequently occurring value, may be preferable. To indicate a range of values, the *variance* or its square root, the *standard deviation*, is employed. When the standard deviation is expressed as a percentage of the mean, the result is known as the *coefficient of variation*. When observations are of the presence or absence of some attribute, then the proportion of those that possess this attribute is normally needed.

Another measurement of deviation is the statement of *confidence limits*, which allows a worker, on the evidence of a sample, to predict the likelihood that the value of a given attribute, at a site randomly selected from the whole population, will fall within specified limits of deviation from the mean. This is obtainable from formulas relating sample size to its mean and standard deviation. Confidence limits are usually established at the 95, 99, and 99.9% probability levels.

It is sometimes necessary to test whether a set of sample values of the same attribute from two different phytogeomorphic units differ enough for it to be worth distinguishing them. For this, it is necessary to apply a test of significance, which calculates from the means, standard deviations, and sample sizes of the two samples the probability with which the difference between the two means could have occurred by chance. Tests of significance on the proportions of attributes present or absent in samples are usually done by the *chi-square test*. Tests of difference between means of measured properties are done by *analysis of variance* or, if there are only two classes, by the *Student's t test*. Both of these assume normal distributions, but where there is serious departure from this, modified tests, such as the *Mann–Whitney U-test* for two classes and the *Kruskal–Wallis test* for more than two, can be used.

An alternative to the study and testing of predetermined classifications of data is the use of numerical taxonomy and multivariate classification. This approach assumes that the site samples are drawn from a continuously varying population or continuum and that the classes into which they should be assigned are best determined by ordination on the basis of maximizing degrees of affinity between groups of attributes. This approach has used *principal components analysis* and *factor analysis*. These techniques proceed by placing the sites in multidimensional space, with each dimension representing a measured attribute. Classes are determined by the clustering of sites within this space, and calculations are made to analyze this clustering to give a minimum number of composite factors that explain the maximum proportions of the observed intersite variations.

Appropriate references on statistical methods are those by Sneath and Sokal (1974) on general applications; Snedecor and Cochran (1978) relating to agriculture; Gregory (1978) and King (1969) to geography; Miller and Kahn (1962) and Krumbein and Graybill (1965) to geology; Greig-Smith (1964) and Piélou (1977) to ecology; and Webster (1977) to soil science.

9.3 FIELD OBSERVATIONS

9.3.1 Introduction

Field observations are based on purposively selected sites or systematically or statistically located field samples, the spacing and distribution of which have been decided according to the predetermined objectives and principles outlined previously. This includes assessing the survey area visually on maps and remote sensing imagery in order to plan the fieldwork and its associated traverses. Sometimes purposive selection of samples are favored because of difficulties of access in rugged terrain or through dense vegetation and the high associated cost or excessive nonproductive working time. Each sample can be regarded as representing all the land up to the midpoint to the next sample in any direction.

It is often useful to have field proformas for recording information because they provide a standard format for records and provide against omissions. Some workers enter their data directly onto punched cards that can then be automatically processed on return, thus avoiding the costs and errors of transcription.

The observations to be recorded fall into three major groups: (a) geomorphic data on landscape, mesorelief, ground surface, and microrelief; (b) vegetal data on plant physiognomy, species, cover, age classes, and so on; (c) land use, and (d) soils and earth materials including soil moisture, water table if near the surface, and presence or absence of hardpan. Samples of the soil and plant species will also probably be collected.

The production of the final reports and accompanying maps following the completion of the collection of field data and later analytical work must be borne in mind. These products require that they be presented in a way that is suitable for the intended readers as well as presented attractively. The report on an area being considered for development should generally include the following elements: (a) an explanation of how it and the maps can be used; (b) a general description of the survey area; (c) descriptions of the mapping units, supplemented by tables and analyses; (d) predictions of vegetal response under varied conditions of climate, natural hazards, and types of management; and (e) a practical consideration of the problems of developing the area. An example of a procedure used in preparing a report is given in Figure 9.4.

9.3.2 Geomorphic Data

Landscape and mesorelief data may be obtained from maps, mosaics, and stereoscopic pairs of aerial photographs, but unless the mapping is at very small scale (1/1,000,000) it must be supplemented by fieldwork. The main aspects requiring observation can be classified under the headings of relief, processes, and landscape esthetics. The study of relief includes an interpretation of the underlying geology and geomorphic evolution based on height amplitude and gradients in relation to the drainage network and the nature of the rocks. Because of the natural complexity of landscape, it is important to record such observations with the aid of field sketches and photographs. The important processes are those that are influencing the landscape within an appreciable time scale such as mass movements, sheet wash, gullying, or the effects of wind action, and their intensity and geographical distribution. For land use planning, consideration of the visual qualities of the landscape should also be included. Such assessments are inevitably purposive, but they achieve division of the area into separable landscape units.

Ground surface observations include microrelief and the nature of surface materials. Microrelief includes the shape and distribution of features less than a meter in height or depth and not more than a few meters in diameter, which are too small to appear on normal maps. Descriptions should be expressed in terms of the detailed genesis and morphology and should include the size, shape, and areal distribution of small features such as rills, hummocks, pits, sink holes, and small dunes.

The character of surface materials includes both their composition and their appearance from above. Where the surface is unvegetated or lightly vegetated it is necessary to assess the proportion covered by stones, sand, silt or clay, and the characteristics of these materials such as the alignments of stones, and the crusting and cracking of silt and clay.

9.3.3 Vegetal Data

The type of vegetal data collected will depend greatly on the objectives of the survey, the intensity of the survey (exploratory, reconnaissance, extensive, intensive) and whether the survey is concerned primarily with the collection of physiognomic data, floristic data, or data from a forest inventory. The last is mainly outside the present discussion, but does involve recording a limited amount of floristic data (e.g., marketable tree species and stand parameters such as stand height, basal area per hectare, and density of the forest cover). In general foresters have neglected, particularly in the tropics, to use geomorphic information.

Before beginning fieldwork or the collection of vegetal data, it is important to examine all available literature, including a study of planimetric, topographic, and thematic maps in order to build up a historical record of the survey area. Large-scale maps, if available, may provide important information on vegetal succession and changes in land use and should be scrutinized with great care. The field-worker should also become thoroughly familiar with the flora of the geographic region, the vernacular and botanical names of species, and be able to identify dominant species when not in flower, to know how to use botanical field keys, and how to collect, dry, and preserve floristic samples in a botanical press for herbarium identification. It is often useful or even essential to establish a temporary herbarium against which newly collected samples of plant species can be checked.

The next step would be a visual assessment of the vegetation types as portrayed on Landsat imagery, mosaics, SLAR imagery, or aerial photographs if these are available. A comparison of old with new aerial photographs will provide valuable information on vegetation succession and changes in land use. The objective is to describe the major vegetal types and to assess the effect of land use changes within the survey area. This includes the identification of the major boundaries of the vegetation types, and with the aid of aerial photographic mosaics and stereoscopic pairs of aerial photographs, the plant cover and stand characteristics. It is only by stereoscopic examination that the major structural differences in the vegetation types can be appraised. On this basis the boundaries of the field sampling strata are established. Obviously the age and season of the aerial photographs or satellite imagery are important to their effective use in the field studies of vegetation.

Beyond this initial deductive approach to vegetation studies, an inductive approach is strongly recommended. At the commencement of fieldwork, the study should always begin with what is observed and not attempt to establish preconceived ideas of the boundaries of the vegetation types–plant communities. The worker should be guided to the collection of field data by the location of the random sample plots or transects or, if this is not possible, then by the location of systematic samples or samples located as objectively as possible on maps, mosaics, or aerial photographs in advance of proceeding to the field.

Workers should be competent in the use of a chain and military prismatic compass, a soil auger, an Abney level, or other hypsometer suited for measuring tree height and slope of the ground and be able to interpret aerial photographs and satellite imagery. They must be able to locate their sample plots precisely on maps, mosaics, or aerial photographs, to record accurately in a field notebook what they observe (species and stand characteristics), and not be influenced by preconceived notions.

In addition, through training and experience workers should be able to evaluate the site quality in terms of its vegetation and conclude whether the vegetation is climax or seral. They should not only be able to observe the phytocenose but also to identify the phytogeomorphic unit of which the plant community is a part. As pointed out by Küchler (1967), in part the landscape determines the character of the vegetation and in part the character of the landscape can be deduced from the vegetation, because the interrelations between the two are most intimate. For further information on the mapping of vegetation reference can be made to Küchler (1967).

9.3.4   Soil and Water Data

When required, soil data can be collected during the collection of phytogeomorphic data. The major landforms and the structure of the vegetation can be used to determine the sites for collection of soil data.

Soil profiles are recorded by the conventional pedological methods (Soil Survey Staff, 1951; Clarke, 1957). The basis for this is the systematic recording of profiles by horizons. Each horizon is assessed for texture by the feel of moistened soil and then observed visually to determine the abundance, lithology, and distribution of stones, the Munsell color, moisture, structure, consistency, cracks, visible salts, or plant roots, and features such as pans (hard layers) or crotovinas (former animal burrows now filled with soil).

The soil profile data are obtained from pits or auger borings made to a depth adequate to penetrate any horizons exploited by plant roots. This depth depends on surface materials and vegetation type, but can normally be taken as 1 to 1.5 m, with occasional penetrations by borings up to 5 m where deeper subsoil materials or a groundwater table are especially important. Borings beyond about 1.5 m are time consuming and increasingly costly to make.

Soil samples are taken for laboratory analysis from representative horizons to check the field horizons. These may either be of disturbed or of undisturbed soil. The former may be obtained by scooping material from the profile face in a pit, or cutting, or by saving layers brought up from auger holes. Undisturbed samples are obtained by driving metal tubes or a coring drill into the ground surface or steps at predetermined depths in soil pits.

Certain tests of land properties relevant to plant growth are best conducted in the field. The most important of these are (a) pH and (b) soil hydraulic conductivity. The former can be determined with a field test kit using color indicator solutions. This is accurate to somewhat better than 0.5 of a pH unit. Greater accuracy than this can only be obtained by sampling for laboratory determination. Water samples are taken of shallow groundwater wherever it appears. Soil moisture is normally measured by oven drying and weighing in the laboratory, but in the field, when repetitive measurements are required, the normal procedure is to use a neutron probe.

Hydraulic conductivity is the intrinsic capacity of the soil to transmit water. This is most accurately measured below the water table under conditions in which flow is laminar and the gravity factor can be eliminated. A suitable method is the pump-out auger hole method of van Beers (1958). The field procedure is to bale groundwater rapidly from an auger hole and measure its rate of rise with an electrical probe and stopwatch and then to calculate the hydraulic conductivity from a formula that relates these values to the depth and radius of the auger hole.

Where the water table is below augering depth, that is, below about 5 m, it is not possible to make a simple field measurement of hydraulic conductivity, and it is necessary to substitute empirical measurements of the surface *permeability* or *infiltration rate* that can be calibrated against hydraulic conductivity measurements where these are available (Hunting Technical Services, 1963; Mitchell, 1973).

In arid regions the most critical factors for plant growth, apart from water, are salinity and alkalinity. Salinity is measured in terms of electrical conductivity and alkalinity by pH, or if land capability is being considered, by the percentage of the exchange complex occupied by sodium ions, usually expressed in terms of the sodium absorption ratio. Methods are given by Black (1965) and the U.S. Department of Agriculture (1954). Water samples are usually analyzed for pH, for total dissolved solids (from the electroconductivity), and for the main constituent cations. These give an indication of the degree and nature of mineralization of the water and the effects on soils and plants.

TEN

THEMATIC MAPPING

10.1 INTRODUCTION

Maps are usually the end product of phytogeomorphic studies and are the basic medium of communication in landscape studies because of their concern with the distribution of visible features on the earth's surface at scales amenable to cartographic treatment. Recent studies have witnessed not only an increase in automated means of data processing, presentation, and display, but also a comparable increase in the number and variety of topographic, planimetric, and special-purpose or thematic maps. Even a list of those associated with phytogeomorphic features is long (Table 10.1).

It is probable that the proliferation of maps at large or medium scales has increased rather than decreased the need for small-scale reconnaissance mapping such as that which satellite imagery provides at scales of 1:200,000 or smaller. There appear to be three reasons for this. First, large-scale maps will only cover limited areas and it may be economically necessary to generalize outward from these to much larger areas, even to countrywide or international scales. Second, there is a need to map new types of thematic data over wide areas for which less data may be available. An example of this is the FAO–UNESCO map of soil degradation at a scale of 1:5,000,000, which follows principles of assessment and local studies at larger scales (1981). Third, there is an increasing demand for syntheses of complex local studies for broad-scale planning purposes. Satellite imagery, particularly Landsat, is now widely

Table 10.1 Selected Types of Thematic Maps Associated with Phytogeomorphic Features

A. General	B. Applied	C. Integrated

Climatic
Single feature
Synoptic weather
Climatic zones
	Agricultural
	Land use
	Land capability—
	  potential land use
	  (general)
		Land unit-land system
		
		Parametric

Hydrological
Water table depth
Rain acceptance
Hydrogeological
Geohydrochemical
	Land suitability—
	  potential land use
	  (particular crops)
	Hazards(e.g., erosion)
	Agroclimatic
	Agroecological	

Geological
Stratigraphic
Metamorphic
Tectonic
Structural
	Soil degradation
	Potential carrying
	  capacity
	  (e.g., human)
	
	Pastoral	

Geomorphic Morphological Morphometric Topomorphic Morphogenetic Morphochronological Physiographic diagrams	Rangeland types Range values Forestry Forest types Timber classes Logging route Working circles cutting series, etc. Site quality	
Pedological Soil survey		
Vegetation Static Succession Type Dominant Communities Nodal	Applied geological Mineral Metallogenic Engineering Surface materials Engineering geological Engineering soils Flood risk Recreational Regional planning Communications	

143

used for summary presentations of earth resource information covered by more detailed thematic mapping.

The basic problem of all landscape mapping is to represent dynamic three-dimensional systems of continuously interacting fluids, materials, and organisms on a static two-dimensional surface. There are many different ways in which this can be done and each map is usually a highly selective picture of nature at one point in time, although maps occasionally portray the changing conditions (e.g., morphogenesis, vegetal succession). The first step is to devise a legend by selecting the factors to be portrayed and the relative emphasis to be given to each. The second step is to decide on the scale, which should be large enough to show all necessary detail but small enough to avoid any unnecessary proliferation of map sheets. Commonly, the field data plotted in map form are twice the scale of the final printed map. This requirement for scale reduction must be considered when planning the field survey and will influence the details that are collected during the field survey. Fairly frequently excessive detail is collected in the field, which is wasteful of time because it cannot all be presented on the final printed thematic map.

Mapping units may be either *specific* or *generic* and either *uniform* or *nodal*.

Specific units or regions are unique and the map that shows them gives no information that can be applied to other areas. Generic units on the other hand are recurrent and allow analogies to be drawn between different areas. For example, Cape Cod (United States), the Cotswold Hills (United Kingdom), Milford Sound (New Zealand), or the Grand Canyon are specific regions, but only if the maps show that they are, *inter-alia*, respectively a peninsula of morainic materials, a limestone cuesta, a drowned valley, and a meandering river very deeply incised into horizontal Paleozoic and later sediments can they be called generic and compared with other analogous regions.

Furthermore most thematic maps show uniform regions with discrete boundaries that indicate homogeneous areas. Dickinson (1930) and Christaller (1966), however, evolved the idea of the nodal region based on interconnections between a central place and its surrounding countryside and, as mentioned earlier in mapping vegetation, these represent classificatory and ordination techniques. The nodal concept has come to be applied to mapping units, which are a functional organization of unlike properties. Although originally relating to urban geography, the concept has application in phytogeomorphology where mapped areas reflect the diffusion of say, water flow, sediment, or plant communities from a central source or node. In floristic studies nodal analysis is used to determine the coincidences existing between species and habitat (Lambert and Williams, 1962).

Where the classification and legend are parametric it depends on operationally defined limiting grades within selected individual attributes. Parametric maps will, for instance, show land with slopes of 0-1°, 1-2°, 2-5°, 5-10°, and so on, or plant communities with percentages of the cover classes of ground vegetation or tree cover (0-10, 10-20, 20-40%, etc.). They are suited to computer manipulation and calculations of variance, but are often less obviously related to the most visible characteristics of the landscape.

10.2 MAP LEGEND

The legend is the key to the map and shows how its content is organized. Most legends have traditionally been of hierarchical type in which the smaller mapping units are grouped into two or more higher categoric levels.

Geomorphological maps normally show the smaller morphometric features of the landscape such as cliffs, breaks of slope, slump features, springs, and stream channels with separated graphic or diagramatic symbols. These are frequently included within larger groupings on the basis of surface materials or forms, or on their processes of formation so that the major types of rock are combined with the landform they underlie to delimit features such as limestone tablelands, river terraces, or active dunes. On the broadest scale these will often be grouped according to geological age, so that on the map separate broad hues will be retained for areas of Paleozoic, Mesozoic, Tertiary, and Quaternary origin.

Vegetation is usually so complex that few of its characteristics can be represented on the map. Often the mapping unit is represented by the vernacular or botanical name of the commonest or dominant species, or the names of the codominant species, and from this it is assumed that the physiognomic or floristic characteristics of the community can be construed. Similarly, foresters often represent the vegetation units by the name of the most economically valuable tree species, which may not be the commonest or dominant one. Woody vegetation, when represented by the dominant tree or bush species, is frequently subdivided by naming the commonest understory species or vegetation type, which in the case of woodland or forest may be a shrub or herb species.

The cartographic presentation in geomorphological maps is based on relatively unobtrusive color washes for surface materials over which are superimposed black symbols for topographic and blue symbols for hydrological features. There is no standard code of colors for different types of material but it is common to choose shades that suggest their origin or their appearance in the field. Warm red or purple colors are often used for igneous or metamorphic rocks, brown for sandstones, yellow or cream for limestones, gray for shales, and black for organic materials. Drift and alluvium are shown in paler pastel shades. Topographic features are overprinted on these color washes in graphic symbols whose boldness is related to their importance in the landscape. For instance, strong lines are used for mountain crests. toothed lines for escarpments, fan shapes for detrital cones and fans, nested lobes for mud flows, star shapes and sinuous lines for various forms of dunes, circles with external radiations for volcanoes and with internal radiations for depressions such as sink holes and dolines. Water features range from pecked lines for fingertip streams to washes for rivers and lakes.

Whenever possible, vegetation maps should commence with the physiognomy and then proceed to its floristic characteristics. Thus the plant formations may be initially shown on the map in primary colors representing forest, woodland, shrubland, and grassland; and if necessary these may be subdivided into areas where intermediate colors of lighter or darker shading show wooded grassland, grassland with scattered shrubs, and so on. Hatching or numbering can be introduced to show (a) the height classes of the vegetation and (b) the density or crown cover of the vegetation. Often on forest maps the stands are divided into age classes including those that are mature or approaching maturity and regrowth classes (saplings, poles, and sawlog size). Symbols are often added to indicate the dominant species. Sometimes a concise supplementary test is added to the margin of the map, which clearly describes each of the vegetation types of the legend and avoids the need to refer to reports and other texts (e.g., Gaussen, 1948; Braun-Blanquet, 1947). Vegetation maps may also contain other supplementary information that, although not descriptive of the vegetation per se, is closely associated with it. Thus Gaussen on his 1:200,000 maps of France provides inset maps at 1:250,000 showing soil types, land use, and climate (precipitation and temperature). Braun–Blanquet on his 1:20,000 maps of the plant community lists the soil types and broadly indicates the potential land uses.

The main alternatives to hierarchical legends are coordinate and multivariate schemes using numerical taxonomy. *Coordinate systems* are parametric and use a limited number of environmental characters to produce a closed legend. In this system a few selected properties are quantified and subdivided into grades, and the mapping classes are based on the combinations of those that are found to occur. In the California Vegetation–Soil Survey, different combinations of vegetation type, age, and stand density are combined into a fractional legend (Küchler, 1967).

Multivariate systems assume that the subdivisions of the environment are polythetic individuals in the form of sites that possess a number of properties. Each property is viewed as a dimension, and the individual is placed in multidimensional space according to its grading in each property. When this has been done for all the individuals, they will tend to form clusters that can be isolated and used as indicators of the groupings to be used as the mapping units. Cuanalo and Webster (1970) carried this further by using principal components analysis to resolve the pattern created by several variables into two or three factors that could then form the basis for the mapping units.

The coordinate and multivariate systems have three clear advantages. First, they make an objective linkage between sites, which eliminates the dangers of assuming hierarchical relationships. Second, they permit affinities to be seen between sites that might not otherwise appear. Third, they introduce computation that both adds precision and reduces labor. The disadvantages are that all attributes must be given equal weight even if they are unequal in importance. Those that are both indicative of genesis and important to land use are equated with others of relatively little importance to either, except in very exceptional cases. Little work of this nature has, however, been undertaken within geomorphic units, and in the future the trend may be toward the use of comprehensive hierarchical frameworks in combination with numerical methods.

10.3 BOUNDARIES

Thematic map drawing is essentially a question of identifying and delineating the boundaries of relatively homogeneous classes. It is common for each mapping unit to have a core area or *node* surrounded by transitional zones or *ecotones* which may on occasion be larger than the nodes. The smallest mapping unit is influenced by the scale and the minimum thickness of drawn lines. The latter is usually between about 0.2 and 0.5 mm, which at a scale of 1 : 100,000 represents a band 20–50 m wide. The smallest practicable mapping unit is a square about 3 mm on a side or a longer strip 2 mm wide, which would represent minimum areas of at least 9 ha at this scale.

In view of the difficulty of defining boundaries in an area of continuous variation it is necessary to accept a degree of impurity within mapping units. A common practice in the earth sciences is to accept that up to 15% of any mapping unit may not represent the class for which it is named; the U.S. Department of Agriculture additionally requires that for soils maps this be reduced to 10% where only a single contrasting area is included. Frequently vegetal boundaries are more easily mapped than edaphic boundaries because the transition can be observed in the field and is associated with a change in the physiognomy of the vegetation or the presence or absence of indicator species or a change in the percentages of dominant or codominant species.

Boundaries are based on a number of criteria. When they surround natural units they overcome the problem of harmonizing a multiplicity of definitive criteria that vary in different degrees and in different places. Such harmony is usually achieved somewhat arbitrarily by using a single dominating criterion in one place and a different one in another. The upper limit of an alluvial fan, for instance, often ends against a steep mountain front where it may be defined on the basis of the sharp increase in slope and the change to solid rock and in the physiognomy and floristics of the vegetation. The lower limit, however, often merges gradually without strong surface change within a broad (plant) ecotone into a detrital plain or the more stratified alluvium of a river valley; thus here the criterion for boundary determination may be a medial line between the two phytogeomorphic nodes, approximately defining the changes in soil texture or the degree of stratification of the profiles.

Parametric maps can sometimes be valuable. These maps preselect a land attribute, determine its values or ranges of values that are significant for a required purpose, and then outline these. Because landscapes are defined by a number of attributes, their parametric mapping requires the overlaying of a set of relevant maps to give composite units. For example, a parametric map of potential land capability could be based on the overlaying of separate maps showing numerical classes of land gradient, soil texture, and of each of the main soil nutrients. The boundaries of different attributes seldom coincide, and the greater their number the harder is the task of harmonizing them into a single boundary line.

There are two methods used for resolving this problem. Maull's girdle method (Grigg, 1967) proceeds by drawing separate maps for the boundaries of each property, superimposing these on a single map, and then selecting the line of greatest coincidence. This method can be refined by drawing in a calculated mean boundary line or by incorporating calculations of the sizes of overlapping areas multiplied by the number of properties involved in the overlap. But although these may be useful compromises amenable to computer-automated solutions, it is preferable where possible in compiling a compound parametric map to resolve such differences of criteria by adjusting the boundaries and if necessary the legend by further reference to the physical factors involved. In forest surveys the boundaries and areas of each volume stratum or marketable species strata usually require adjustment after the statistical analysis of the collected field data.

A second method is the statistical *principal components method*, first suggested by Hagood et al. (1941), and the associated technique of *direct factor analysis* that has been developed considerably since then (Cole and King, 1968). The principle of this method is to simplify the problem of relating a number of mapped attributes to each other by using multiple correlation in order to find ways of grouping them into sets that are more or less similar.

The correlation of two attributes can be shown by a regression line on a graph. The correlation of three is a three-dimensional planar or trend surface that can be represented rather clumsily on two-dimensional paper using perspective. The correlation of four or more attributes requires multidimensional space that cannot be represented graphically and mathematical calculations that are prohibitively long unless processed by computer. Factor analysis was derived to meet this need and has been applied to terrain study by Cadigan et al. (1972). A computer carries out multiple correlation analyses in such a way that the resulting values form clusters in multidimensional space. These clusters can be interpreted in terms of a small number of trends called *principle components*. These are not themselves the attributes, but represent contributions of various sizes from a number of the attributes and explain part of the variations between them. The amount of explanation contributed by each principal component is quantified as its *eigenvalue*. This term can be defined as the value of each characteristic or latent root of a correlation matrix. It may be thought of roughly as the length of vector which passes through a scatter of points in multidimensional space so that it reduces to a minimum the distance between the points and the vector. Two or three eigenvalues normally explain most of the variations within the matrix, and these, as an expression of the principal components rather than the original land attributes themselves, can be used as the basis for mapping boundaries.

10.4 THEMATIC MAP PREPARATION

Whenever possible, existing planimetric or topographic maps should be used to provide the base on which the thematic data are plotted. Planimetrically controlled or semi-controlled photomosaics may also be used. These may require some ground survey coupled with aerial photographic interpretation (Figure 10.1). Many so-called thematic maps are more correctly termed *sketch maps* because they do not have the necessary mapping accuracy for their scale.

Planimetric maps show only the horizontal positions of ground features. They are distinguished from topographic maps by the omission of relief in measurable form and include no contours. When produced from aerial photographs, they are adjusted in relation to ground control points so as to eliminate radial distortions. Topographic mapping requires in addition a network of bench marks to indicate spot heights, and contour drawing additionally requires skilled photogrammetric interpretation in plotting machines.

The final scale selected must clearly take into consideration the purpose of the survey and who will be the eventual user, but it is usually also related to the type of available aerial photographs, although with the newest aerial photography using improved lenses, filters, and film quality, the map scale may be as much as twice the aerial photographic scale.

Thus in preparing both geomorphic and vegetation maps, aerial photographs at 1:50,000 acquired with modern equipment may be enlarged to produce planimetric maps at 1:25,000. It must be borne in mind, however, as mentioned earlier, that it is not practicable to outline on the final map an area smaller than about 3×3 mm or a strip less than 2 mm in width. If it is necessary to reduce the scale of the map for final production, and this requirement is overlooked, the operation of combining many of the units into larger ones may

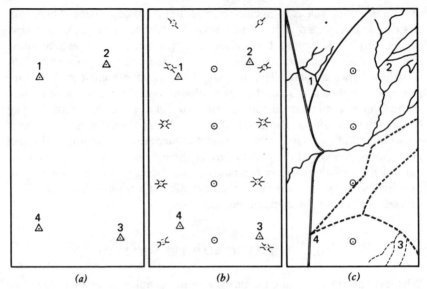

Figure 10.1. Preparation of a thematic map. (*a*) Ground control points △ plotted on the map base and obtained either by field survey or from existing planimetric or topographic maps at the same or larger scale. (*b*) Photographic control points ⊙ obtained from stereoscopic pairs of aerial photographs and transferred to the map base to supplement the ground control points. (*c*) Major natural features (e.g., drainage lines) and manmade features (e.g., roads) are transferred to the base map. Not shown: in combination with field checking, aerial photographic information on the geomorphology and vegetation is added to complete the thematic map.

be tedious, costly, and result in the loss of important information or the introduction of mapping error associated with the simplified legend.

Cartographic accuracy is also important, but the concepts are different for the base maps and for the thematic information that is shown on them. Since the 1940s the convention has been that *horizontal accuracy* on planimetric or topographic maps is assured when, at scales larger than 1:20,000, not more than 10% of well-defined listed points are in error by more than $\frac{1}{30}$ in. (0.85 mm) and, at scales smaller than 1:20,000, by more than $\frac{1}{50}$ in. (0.51 mm). *Vertical accuracy* on topographic maps is assured when not more than 10% of elevations tested are in error by more than half the contour interval. More recently, attempts have been made to derive a formula based on an allowable standard error d, defined as

$$d = \sqrt{\Sigma e^2/(n - 1)}$$

where e is the error and n the number of observations and the accuracy is expressed in terms that are linearly related to map scale (Robinson, et al., 1978).

Accuracy in thematic mapping is, however, less concerned with the precision of individual map positions than with the general truthfulness of the phenomena being portrayed. In being more relational the map may sometimes have to use symbolism at the expense of the strictest cartographic exactness as, for instance, when boundaries to important properties have to be generalized across sites or areas that may have doubtful or aberrant values.

If existing maps are available at the same or larger scale than that of the final phytogeomorphic maps, there is no difficulty in providing the control to give such accuracy. If no suitable base maps exist or if the scale of those available is considerably smaller than that required for final production, the only alternatives are either to establish expensive ground control or else to accept some loss of accuracy and portray geomorphic and vegetation features in diagrammatic or sketch form that gives information on their reliability.

Unless computer-assisted techniques are used, the usual procedure in preparing the thematic map showing geomorphic and vegetation data is to transfer the results of fieldwork onto stereoscopic pairs of aerial photographs or photographic mosaics.

The data may be transferred from the photographs onto the base map via a transparent overlay on which ground control points have been scribed and into which the photographic features can be fitted. Alternatively, the photographs may be assembled directly as an uncontrolled, semicontrolled, or controlled mosaic and separate overlays prepared for geomorphic and vegetal data. Place names and similar information are annotated onto similar overlays, and these are combined during printing to provide the final black-and-white or colored map.

It is important to obtain a legend that is as clear and comprehensive as possible with a minimum of complication. Different styles of geomorphic mapping exist and legends tend to be complex because of the need to combine information on the nature and age of surface materials with a representation of the landforms within a legible and graphic format. The size of the problem can be seen from the fact that the scheme used in the USSR, which gives attractive maps and is about the most comprehensive, has over 500 items in the legend but yet lacks slope data (St. Onge, 1968). It is common practice to adopt a simpler scheme with a shorter legend (e.g., Verstappen and Van Zuidam, 1968). Vegetation is usually so complex that only a few characteristics can be mapped directly. Therefore, vegetation maps often show only classes or categories. Vernacular or botanical names may be omitted but are associated with symbols included in the legend. For example, density and height classes are usually reduced to the fewest essential groups. If communities are to be represented by species names or by physiognomy, only dominant species are used and then only in the form of a symbol. Life-forms are restricted to a few major classes such as woodland, shrubs, and grassland. Geomorphic and vegetal

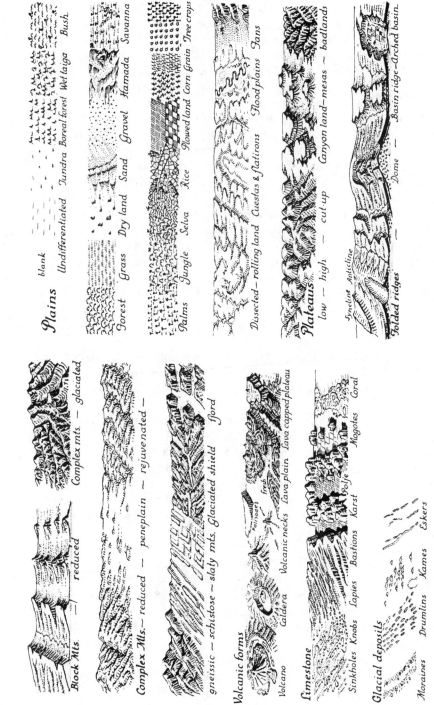

Figure 10.2. Thematic symbols that can be used in graphic mapping and block diagrams (Raisz, 1962).

data may be combined on the same map by establishing first the boundaries of the land units and then symbolizing within each the phytogeomorphic data.

10.5 PRESENTATION

It is most important that a thematic map should be clear and make a good visual impact. They are usually printed on, or controlled by, topographic base maps showing grid coordinates, height contours, and gazetteer information. Thematic data are added in three main styles which may be either separately or in combination: (a) area shading, or coloring, (b) graphic description and symbolism, or (c) the use of annotated aerial or satellite imagery. Area shading or coloring is the conventional method used in most thematic mapping, and can be performed digitally by computer when the track of the boundary lines in terms of x-y coordinates is established.

The units are shaded or colored in, using conventions likely to be meaningful to the user such as blue for water and green for vegetation. To improve clarity or artistic effect, it is sometimes desirable to use graphic forms of presentation such as topomorphic maps (Marchesini and Pistolesi, 1964), block diagrams (Figures 3.2-3.9), ground level panoramas, or stereoscopic profiles (Figure 8.8). Where possible it is desirable to add symbols suggestive of particular rocks or vegetation types (Figure 10.2). It is sometimes both cheaper and more graphic to use satellite imagery or aerial photographic mosaics or orthophotomaps reproduced in halftone as the background on which boundaries are drawn (Figure 8.3).

10.6 TYPES OF THEMATIC MAPS ASSOCIATED WITH PHYTOGEOMORPHIC STUDIES

A list of some types of maps relevant to phytogeomorphic studies is given in Table 10.1. Although they can be classified into general and applied types, there is clearly a considerable overlap between them. A general geohydrochemical map will, for instance, emphasize attributes important to human, and applied land capability or land suitability maps will include much natural environmental data. The general maps can be subdivided according to scientific discipline and the applied maps according to the type of user. The overlap within each group is likewise considerable.

Detailed integration of diverse types of thematic mapping is not easy and not always needed, but there are a number of arguments favoring the achievement of a broad correlative framework. Earlier chapters have emphasized the integrated survey approach based on land units of varying magnitude. The essential principle has been to map these units at a relatively small scale and to accompany the map with a comprehensive body of data that explains the principles on which each may be best subdivided and the detailed character and practical capability of each of their subdivisions. Combining resource surveys onto a single base in this way makes for economy of effort and reduced cost. It enables workers from different disciplines to use the same base maps and remote sensing imagery and to exchange ideas while the survey is being conducted. The result is to give a map that usefully combines the basic data on natural resources and shows the interrelations between them.

There has been little effort worldwide at mapping phytogeomorphic data at larger national mapping scales, and even the separate mapping of geomorphic data and vegetal data (other than as forest maps) is poorly represented internationally. Phytogeomorphic studies therefore have to rely principally on topographic, geologic, soil, economic forest type, and land use maps, and on the direct analysis of aerial photographs and the imagery of earth resources satellites at very small scales. The effects of small-scale map generalization including broad climatic zonation lead to an increased coincidence of landform and vegetation boundaries. Harper (1943), for example, used geological data for mapping the main forest types of Alabama at a scale of 1:4,000,000 and was able to show the close correlation between geological features and vegetation. Under favorable conditions similar correlations have been demonstrated at larger scales (1:1,000,000), but the emphasis placed in geological maps on stratigraphy rather than lithology often makes it difficult to discern the effect of rock type on landforms and soil development.

Climatic maps containing information useful in phytogeomorphic studies are broadly divisible into two main types. First, there are those that show single factors, either as numerical values or coded symbols at points of observation or as areas defined by isobars, isohyets, isotherms, and so on. Second, there are those that synthesize a number of factors into periodic synoptic weather maps or broad climatic classifications. The latter follow those of Köppen (1923) and others, but are now applied in more detail to single climatic zones such as the UNESCO map of arid regions (1979) or to selected regions such as the climatic maps of parts of Africa by the German Research Society (1977).

Geomorphological maps have developed from geological maps, that have traditionally been based on the stratigraphy of solid and drift deposits, although specialized maps show features such as tectonics (UNESCO, 1968), structure (Demek et al., 1972), and metamorphism (UNESCO, 1974). Geomorphological mapping has been reviewed by St. Onge (1968). Although first suggested by Passarge (1919), geomorphic maps only began to be developed in a number of countries after 1950. Because their maps differed considerably, the International Geographical Union in 1960 set up a subcommission on geomorphological mapping to standardize legends and symbols for showing the appearance and lithology (morphology), the dimensions and slope values (morphometry), the origin (morphogeny), and the age (morphochronology) of each form (Demek 1972). Quaternary maps (UNESCO, 1967) can be included among these types. Verstappen and Van Zuidam (1968) produced a simplified practical legend of 469 symbols divided among 12 groups representing the broad lithogenetic types of landscape—volcanic, fluvial, eolian, and so on. This has gained wide acceptance as a basis to which local modifications can be made. The Sheffield School devised a complementary scheme for the morphometric mapping of small landforms (Savigear, 1965). Geomorphological features have also been graphically represented by block diagrams and physiographic panoramas by many authors, among whom Lobeck (1939) and Raisz (1962) are notable, and by the geomap system for topomorphic mapping from aerial photographs (Marchesini and Pistolesi, 1964). The growth of tourism in the past decade has led to a development of colored and annotated oblique views of landscape usually based on aerial or satellite imagery (U.S. Geological Survey, 1970).

The *vegetation* map is a two-dimensional representation of three-dimensional field observations. The terrain in part determines the character of the vegetation and in part the character of the terrain can be deduced from the vegetation. If the intention is to combine vegetal with geomorphic data, it is often most useful to superimpose the vegetation onto a topographic or geomorphic map base, because the geomorphic influence on the vegetation can be more readily inferred.

At the reconnaissance level of mapping (1:250,000–1:1,000,000), emphasis is placed on the general physiognomy of the vegetation at the level of plant formations. As the need for larger scale mapping (1:100,000) increases, the map is likely to contain information on plant subformations, height classes, and density classes. As the demands of management further increase, information is introduced with larger mapping scales (1:10,000–1:50,000) in the form of dominant and codominant species of the plant associations and plant alliances. At very large scales it may be found convenient to map vegetation succession and the plant communities in a continuum by gradients or nodally. Also, some vegetation maps will contain habitat data relevant to the establishment of plant communities.

Hydrological maps have tended to emphasize groundwater in addition to the surface water channel network shown on topographic maps. They have been mainly of three types: groundwater depth, hydrogeological, or geohydrochemical. Groundwater depth maps usually show the subterranean contours of the water table and are used in water supply and drainage studies. Hydrogeological maps add the distribution of the geological formations and their character as aquifers or aquicludes (UNESCO, 1970). They thus tend to be somewhat more complicated and to depend on color for clarity. Because of the relationship between soil type and the rate of rain infiltration to supplement the aquifers, the Soil Survey of England and Wales (1977) has published a winter rainfall acceptance map of the country. *Geohydrochemical* maps show the nature and degree of mineralization of the water (Experimental Cartography Unit, 1970) and their legends for different scales have been standardized internationally by UNESCO (1975). Maps showing water-borne hazards such as flood risk are produced in vulnerable coastal zones and populous river catchments such as those around the North Sea.

Soil maps reflect the natural basis of most present soil taxonomy, begun by Dokuchaiev and his school in the late nineteenth century but developed internationally since then. The main advances in the past two decades have been progress in the refinement of definition and nomenclature of characters, in the greatly increased geographical coverage of detailed soil mapping, and in the growth of international cooperation culminating in the completion of the FAO–UNESCO *Soil Map of the World* (1974).

Applied mapping has served mainly the interests of specialists concerned with agriculture, forestry, engineering (both civil and military), urban planning, mineral exploration, and tourism.

Land use, land capability, and *land suitability* maps have wide uses in agriculture and forestry and may be broadly of three types: physical, agronomic, or economic. Physical maps of land capability classify and grade the topography and soils in accordance with their value for agriculture, pasture, and so on. Agronomic suitability maps assess the land potential either for particular crops such as cocoa (Smyth, 1966) or cocoanut palms (Jenkin and Foale, 1968) for type of land use such as arable land, pasture, or forestry, or generally in terms of agroclimatic potential or agro-ecological zones (FAO, 1978; Dudal et al., 1982). Storie (1964) suggested a numerical index based on physical land attributes by which its potential productivity for any crop could be related to a standard in California; this principle has since been widely applied elsewhere. Economic land suitability maps show areas in terms of their potential economic return for farming at specified management levels. Such a system has been developed, for example, in New York State (Conklin, 1959). For geomorphological hazards Verstappen and Van Zuidam (1968) have suggested a method of *morphoconservation* mapping and FAO–UNESCO have initiated a *Soil Degradation Map of the World* at a scale of 1:5,000,000 (1981).

Land use mapping shows areas occupied by different types of activity and has increased in sophistication and complexity since the pioneer work of Stamp and Willatt's *The Land Utilization Survey of Britain* (1934). On the whole the move has been toward making land-use classifications more comprehensive and hierarchical so that, for instance, cereals, roots, and fodder categories would be subsumed under an arable class, and also toward making classifications show the nature of the economic activity performed on each unit rather than merely the land cover they carry at a given moment, that is, replacing an area-based with an activity-based emphasis.

Engineering maps began with the overprinting of engineering information onto geological or soils maps, but now involve specialized techniques. These can be used for documenting data such as those from boreholes for special purposes such as landslide susceptibility or soil strength; as multipurpose, combining these facts; and as interpretive and diagrammatic, giving cross sections and three-dimensional representations (UNESCO, 1976b).

ELEVEN

DATA PROCESSING
AND STORAGE

11.1 CONCEPT OF A PHYTOGEOMORPHIC
INFORMATION SYSTEM

Many organizations need to store and process location specific data of a phyto-geomorphic nature. These include local and national authorities concerned with rural planning and development, multinational firms with major interests in the environment and international bodies charged with broad-scale assessments of renewable and nonrenewable natural resources.

When viewed at anything more than local scale, the vast amount and variety of environmental data involved and the wide range of locations from which it is derived pose serious problems of management. This has led to a need for intelligence systems to store and manipulate such data in a way that will be readily and rapidly available for planning purposes. Such systems are known broadly as *geographic information systems*. They can be defined as organized groups of activities and procedures involving the collection, storage, manipulation, retrieval, and presentation of geographically referenced data, including phytogeomorphic data; and they are valuable to the extent that they offer an efficient way of handling unwieldy or large quantities of data and information.

A geographic information system consists of these six major subsystems: management, data acquisition, data input and storage, data retrieval and analysis, information output, and information use (UNESCO, 1976a). Management includes the normally recognized management functions that control the operation of the whole system. Data acquisition covers the research aspects by which information is obtained. Data input and storage include the coding, classification, and translation of the raw data into formats, where possible numerical, in such a way that they can be fed into the store and readily retrieved. Data retrieval and analysis is the phase in which they are manipulated and processed to yield useful information. Information output provides material in the form appropriate to the user or implementing authority and information use covers the interface between these and the system as a whole. Computer assistance may be used to some extent at all stages, but especially in data input, storage, retrieval, and analysis. The ready availability of low cost computers and associated software is giving considerable impetus to the development of geographic information systems.

The required output from a geographic information system emphasizing phytogeomorphology includes the following:

1 The display of the distribution of one or more landscape characteristics, such as landforms, natural vegetation types, and land use and associated information (e.g., geological, topographic, climatic).

2 The display of changes over time, such as the conversion of natural forest to farmland, pasture to arable cropland, or rainfed agriculture to irrigated crops, or abandoned farmland to scrub or to regrowth forest or to forest plantations.

3 The combination of selected parameters to predict land potential for particular needs, especially the introduction of new cash crops in response to changes in national development and international demands.

11.2 CRITERIA FOR APPRAISING A PHYTOGEOMORPHIC DATA PROCESSING SYSTEM

In general, in judging an intelligence system, it is important to assess the degree to which information is presented to users in language and format that is not only accurate, concise, and clearly comprehensible but also simply reproducible and graphic to the whole range of those who will use it, most of whom will be nonspecialists. This often favors maximizing the ratio of maps, diagrams, and illustrations to the written text. When the amount of data to be handled is modest and can be presented two dimensionally, manual or simple graphic methods will often suffice and will be most cost effective. Otherwise, a computer-assisted system will be needed.

The following criteria can be used to measure the quality of systems which have been developed to store and process data:

1 The degree of refinement of the locational identification.
2 The amount of data related to each specific location and the amount of data that can be stored.
3 The manipulation and handling of the data in store, including manual and computer processing and output devices.

These can be expressed most effectively in graphical form, representing multidimensional space. Thus in Figure 11.1 the quality of the system is assessed in terms of the three-dimensional distance that a geographic information system lies away from the origin of this space.

Index 1 is the character of the location identifier. It represents the number of geographical points or areas about which data are held and its degree of precision in indicating geographical location. It ranges from standard geographical place names at the simplest through coordinated points to areas whose limits are closely specified. This index is also conditioned by the effectiveness with which individual sites can be grouped into valid natural units for the purposes envisaged. When many data are involved, the accuracy of the coordinates for any point depends on the original survey and on the registration of data from different planes onto a common base.

The data may be based either on x-y coordinates or on grid squares or polygons defined in terms of their limiting coordinates. The x-y coordinate structure allows high precision of geographic location and high storage efficiency, but is difficult to manipulate and analyze areally. The grid system on the other hand allows easier data analysis and manipulation for operations such as overlaying, but it reduces the precision of geographical location, necessitates the inclusion of the natural variations within units, and makes storage more difficult (Townshend, 1981). These variations are minimized when the areas are polygons defined to coincide as far as possible with natural units.

Areal units have been used at a range of scales. Postal systems are sometimes based on a 10-m grid. The U.S. Large Area Crop Inventory Experiment (LACIE) used Landsat cells of 1×1 km (Driggers et al., 1978). Satellite rainfall monitoring in Africa used cells of 50 and 100 km^2, and squares of 7.5×7.5 km and 15×15 km were recommended for a world photographic index (Howard and van Dijk, 1980).

The quality of the information clearly depends on the accuracy of the locational data and on the predictive value of the areal units into which they are grouped. A grid system facilitates data analyses and the manipulation required for overlaying. This is the basis for showing the spatial correspondence

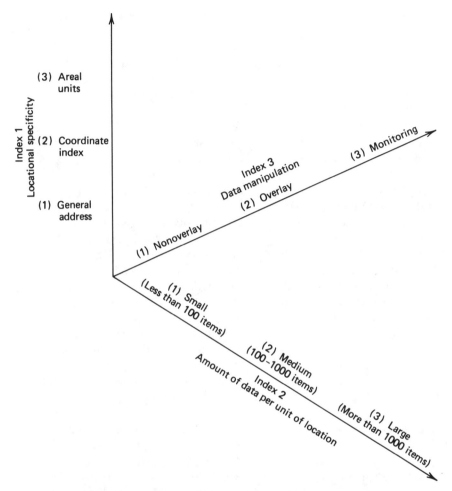

Figure 11.1. Diagram illustrating the parameters for measuring the value of a geographic information system.

of different properties and thus for composite analyses and mapping of multiple data sets over different geographical localities. But it may have the disadvantages of aggregating dissimilar features in the same areal unit and thus reducing the precision of geographical location and creating data that are difficult to store efficiently.

Index 2 relates to the amount, chronological frequency (temporal resolution) and precision of data attached to each location, and the degree of comparability between those from different locations. Examples of organizations with large amounts of data for each point are national census, national meteorological observations, and the World Meteorological Organization's World Weather Watch which has several thousand sequential values for each station. Some organizations include phytogeomorphic data, notably for natural environmental, agricultural, or geological surveys such as the Canadian Geographic Information System, and the EROS Data Center, South Dakota, for information gathered by Landsat.

Index 3, data manipulation, is the capability of the system to store, process, and retrieve the data. The basic facility is a simple indexing to secure its ready availability, but the quality of the system depends on its ability to perform a number of operations of increasing complexity and sophistication. There are basically three levels of this system. First, there are the methods that are simple and do not involve overlaying more than one data plane. These begin with the basic retrieval and increase in complexity through summaries, scale changes, selective searches, distance and area measurements, map generation, calculations of attributes within defined areas, comparisons between areas, route determinations, and automatic contouring, to which may be added such refinements as hill shading or layer shading.

The second level is the introduction of one or more further data planes and the application of the analyses listed above to a combination of them. Finally, the third level occurs when the performance of these operations is backed up by a monitoring capability that follows continuous change and gives signals when required of opportunities and dangers that can trigger automatic remedial measures or call for references to external decision making, such that environmental hazards can be foreseen and countered and planning priorities determined.

It is essentially the combination of the capabilities from these three indices which determines the value of the system. The major subdivisions of each also shown in Figure 9.1 indicate levels of increasing sophistication with distance from the origin and can be viewed cumulatively. Manual methods of analysis will in general suffice within the first subdivision of the indices. At the second level there is a need for introducing automated methods and at the third level almost total reliance will have to be placed on computer storage and retrieval.

11.3 MANUAL METHODS

A manual system of storage of phytogeomorphic data should contain two basic elements that are mutually referenced:

1 Descriptions of land units giving locations, geographic grid coordinates, and phytogeomorphic data in such a way as to facilitate the recognition of analogues between different land areas.

2 Items of practical information of importance to current land use, change in land use, and land evaluation for local or national development.

A simple method is the *uniterm* system using optical coordinate matching of cards. It works as follows. Each item of practical information is analyzed for its content and these are reduced to a number of single terms called *descriptors*. In general these descriptors are of four types: (a) land units, both specific and generic; (b) grid references; (c) information about the form of items, that is, map and photograph; and (d) information about the content of items, such as yield of timber, diseases of wheat, and soil strength for buildings. To take two examples, first, an item of information might be that within defined climatic limits onions grow well on peat soils in reclaimed lacustrine basins along the southern shore of Lake Erie. Onions, peat soils, lacustrine basins in general, and the grid coordinates of the observation are each regarded as descriptors. Second, in some tropical savanna woodlands of central and east Africa, onions grow best in soils of old termitaria that have a thicket type of natural vegetation and are relatively high in calcium. The descriptors include onions, termitaria, soil type, woodland, and thicket.

For each descriptor a *feature card* is prepared. This card has an array of numbered punchable spaces. When an item relates to a certain descriptor, its number is punched on the feature card for that descriptor. The same numbered space would be punched out on the feature cards for onions, peat soils, lacustrine basins, and so on. To find an item relating to two or more descriptors, one superimposes the feature cards on a light table. The places where the light shows through identify the numbers of the relevant item or items.

Feature card retrieval can be manual or mechanical. Where there are relatively few cards, they can be indexed and placed in alphabetical order. When their number exceeds a few hundred, it is simpler to sort them by means of needles inserted into the punched marginal holes.

The item cards are kept separately, and when the store begins to grow it is desirable to separate it into fast access and slow access parts. The former consists essentially of the bank of item cards on microfilm with retrieval, viewing, and printing facilities. The slow access store is the repository of all original material unsuitable for reduction or for which there is limited demand. It includes books, articles, maps, aerial photographs, and bibliographic references to material not held with an indication of where they may be found.

When the material is provided for a user, it will usually be necessary to supplement the printout material by editing it into a suitable format. McNeil (1967) has suggested that output presentations on particular land areas take one of four forms: general briefs, specific briefs, maps, and aerial photographs. General briefs give all available background on an area in generalized form. Specific briefs add information relevant to one type of land use such as the yields of a particular crop or the engineering properties to be expected in the area.

It is also possible to generate provisional homologues about unknown or inaccessible areas by extrapolation from known areas that are recognizably analogous. The procedure is to scan the store for areas that are as similar as possible in terms of environmental data and that also appear similar on remotely sensed imagery. Information subsequently obtained about the new area may then be fed into the store to refine its capability for further interpolation. Maps and aerial photographs can be annotated thematically to define and illustrate phytogeomorphic characteristics of importance.

11.4 COMPUTER-ASSISTED METHODS

When we use large volumes of spatial data of several different kinds, it is necessary to automate the information system. Within the past decade the need to use large volumes of spatial data of several different kinds has led to automated geographic information systems incorporating all stages from data acquisition to information use. Systems with national and international dimensions have been developed for particular topics, such as the topographic factors affecting military operations (Anstey, 1960; Pearson, 1979) and for soils (Bie, 1975; Kloosterman and Dumansky, 1978). Countries are now increasingly developing geographic information systems that contain phytogeomorphic information and in which the stored data, when retrieved, can be developed in thematic map form. In Chile, for example, data from existing thematic maps showing landforms, vegetation types, geological structures, soil types, and planimetric features (e.g., rivers, roads, and spot heights) are digitized and stored for simultaneous or selected retrieval according to their geographical coordinates. An area of 60,000 km^2 covered by a map sheet at 1:50,000 takes about one week to digitize and computer store ready for retrieval.

In principle, remotely sensed satellite data, especially in digital form, can also be incorporated into geographical information systems. One can either incorporate the remotely sensed satellite imagery into the geographic information system pixel by pixel as a separate data plane or else subdivide it into a number of planes, each of which can be combined with components from the imagery so as to improve the overall classification of each plane.

Nevertheless in practice such integration of satellite data presents major difficulties (Knapp and Ryder, 1979; Townshend, 1981). There are a number of reasons for this. First, it is necessary geometrically to correct the imagery so as to register it with the grid coordinates of a suitable map projection (e.g., horizontal Mercator). Second, image quality due to problems of cloud cover, heavy haze, poor tonal contrast, and coarse resolution (40–80 m) may make it less valuable as a data source; and third, the imagery requires specialized interpretation before it can yield information about the landscape.

Townshend (1981) reviewed some of the approaches that have been made to computer-aided geographical information systems incorporating several layers of information. Most contain a considerable component of remote sensing in all layers. Examples are the studies of the distribution of agriculture on soils of different qualities (Cowan et al., 1976), residential and recreational land suitability (Hicks, 1977; ESRI, 1979), temporal changes in land characteristics (Schlesinger et al., 1979), and the assessment of land resources of parts of Indonesia, which have been designated for settlement from more congested districts (Dent, 1980).

At a wider scale, automated geographic information systems covering parts of North America were reviewed by UNESCO (1976a). These included the Canada Geographic Information System (CGIS) and systems developed in San Diego county, California, part of Tennessee, and the states of Minnesota and New York that covered a wide range of environmental and natural resource data. The organizations concerned carried out research and pilot studies which established the possibility of automatically transferring their varied data from graphic to digital format for storage and processing.

A number of studies, some of which are quoted by Townshend (1981), have shown that a stratified subdivision of satellite imagery by interpreters, based on physiographic principles and a general knowledge of ground conditions, may be preferable to digital analyses alone. This is because human analysis can select any part of an image instantaneously and interpret complex patterns (including shadows) on the basis of experience with greater speed and reliability, can separate the types of phytogeomorphic data under different sun angles and conditions of haze, and can probably operate at lower cost than is possible with purely automated means. Nevertheless the rapidly increasing quantities of data from imagery, particularly with higher spatial and temporal resolution of the satellites of the mid-1980s and the growing complexity of data manipulation, will inevitably encourage the expansion of computer-assisted methods.

11.5 DIGITIZING MAPS

Several techniques are available for digitizing maps, including rotating drums and flatbed digitizers. Maps may be digitized, for example, using a specially designed flat plastic table within which there is a printed grid of fine parallel wires. The map to be digitized is placed on the plotting table and a manually operated electrical cursor is then used to follow map lines and automatically to encode onto magnetic tapes or floppy discs the x-y coordinates of the grid wires that are crossed by the cursor. Headings and labels are added by a microprocessor and the whole transferred to magnetic tapes. This is then usually processed by microcomputer, which normally has a cathode-ray display permitting visual checking.

Also, uncoded thematic map data can be scribed automatically onto a flat-bed plotter either with a pen or by a light spot projector shining through a rotating disc. This has an array of symbols, including a choice of lines of different width and pecking length and pattern. Furthermore, to produce thematic maps showing different types of area in different colors, labeling is most conveniently done at the same time as digitization by marking boundaries according to the areas on each side of them. Color masks are assigned to the different areas. Each color is then printed onto a separate digitized sheet before they are all superimposed in final production.

It is usually necessary to convert arbitrary digitizer coordinates to an absolute grid. This can be done by first determining exactly the corners of each map sheet, digitizing these, and then transforming the coordinates of all intervening details by using a software program and the computer to make bilinear interpolation between them. This permits maps originally digitized as separate sheets to be reassembled automatically into their correct relative position. The transformation of coordinates between different projections can be done by the same method.

When it is required to reduce mapping scales, some of the points must be discarded in order to generalize shapes both for graphic appearance and for maintaining speed. The key problem is how to decide on which points to discard. This is done either by retaining every nth point or by retaining only those more than a certain distance from the previous retained point.

For automatic contouring the operator scans a grid of spot heights for each altitude for which a contour is to be drawn. When two adjacent grid values bracket the desired altitude, the position of the contour between the two points is estimated by linear interpolation and stored. When the scan is complete, these interpolated contour positions are arranged into a drawing sequence and joined. Isometric block diagrams can be drawn from x, y, and z values when the orientation, vertical exaggeration, and scale of the diagram are specified.

When lines bounding areas are digitized in the normal way, it is possible to make areal measurements. The areas between each line and the x axis can be calculated by computer and then totaled.

11.6   PROCESSING OF PHYTOGEOMORPHIC DATA

Geomorphic and vegetal data are being incorporated at various levels into data banks by national organizations. The current use of such systems is conditioned both by the amount of numerically referenced data they contain and the sophistication of the processing methods.

The Canadian Geographic Information System (CGIC) includes data on present and potential land uses (UNESCO, 1976a). It entered routine use in 1968 and is in two parts consisting of a data bank and a set of procedures for manipulating the data within it. The bank accepts maps that are digitized onto magnetic tapes as they are scanned on a rotating drum (see Section 11.5), an important factor because there are over 30,000 maps for the entire country. Each scanned strip or raster scene may be only 0.1 mm wide. The system permits measurement of map areas on line lengths and counting of point frequencies. Boundaries are defined by polygons and different types of data within these can be retrieved, compared, and classified. Locations can be identified with specified characteristics. The system is also able to attach a reliability factor to all information that is amenable to continuous modification and updating. The system is especially advanced in its inclusion of techniques relating to the compact storage of boundary data and rapid comparisons between maps.

Some other systems aim more specifically at managing spatial data on land productivity for agriculture that is based on a coordination of environmental with land-use data. The Comprehensive Resource Inventory and Evaluation System (CRIES) of Michigan University (1979) aims to provide an in-country capability to inventory the current and potential production capacity of agricultural resources in a way suitable for transfer to other countries. Essentially the method consists in subdividing the landscape into units at two different orders of magnitude:—larger scale resource planning units (RPUs) and smaller production potential areas (PPAs) on which comprehensive data are held on environmental factors and crop yields under different management levels. These are cross-referenced with administrative areas. The system has two phases—the assembly and computer coding of this data set and linear programming. The output gives crop choices and likely costs and yields for each crop and area.

11.7 GEOGRAPHIC INFORMATION CENTERS

Because geographical information systems exist to deal with large amounts of data, often of considerable complexity, the need has arisen for centers which can manage and process the data. These are known variously as information analysis centers, specialized information centers (AGARD, 1970), or where the scope is worldwide, world data centers (e.g., ICSU, 1979). An important aspect of geographic information systems lies in their capability of relating environmental factors to vegetal productivity by automated means over wide geographical areas.

In the long run geographical information systems appear to hold the key to the solution of many information management problems, and are likely to become the accepted method of operation in the future. They are mainly limited today by their relatively high operating costs, the difficulty of demonstrating their value, and because potential users are often unaware of the services they can provide. Nevertheless this type of center has increased in the past two decades and they now cover many topics both nationally and internationally ranging from relatively narrow fields such as heat transfer and fluid flow in materials (Cousins, 1970) and maritime pollution (Langston, 1970) to wider subjects such as glaciology, vulcanology, and oceanology (ICSU, 1979).

TWELVE

PHYTOGEOMORPHOLOGY IN LAND USE MANAGEMENT AND PLANNING

12.1 INTRODUCTION

In the past 20 years the world population has increased from an estimated 3 billion persons to 4.5 billion and is expected to reach over 6.2 billion by the year 2000 (Salus, 1981). In China, for instance, there are already 300 million people under 30 years of age, and worldwide about 95% of the population increase can be expected to occur in developing countries. Looking ahead, a 50% increase in the world food production may be necessary in the next 20 years if the present standards of food consumption are to be maintained, although already malnutrition and famine are serious in some countries (FAO, 1981). However, if all cultivated land could be used in an optimal way, with unrestricted movements of produce, there would be sufficient food for all (Dudal, et al., 1982). If current agricultural growth of about 2.7% a year can be maintained, this will just keep ahead of the predicted world population growth of 2.3%, but will fall short, for example, of the population growth in Africa (FAO, 1979).

Such statistics clearly indicate that land is a very limited world resource, the use of which needs to be much more carefully planned in the future, because of the rapidly increasing world population and the scarcity of land suitable for some purposes. In fact, only about 25% of the world land resources are suitable for cultivation and these are unfortunately unevenly distributed. The next two decades may prove therefore that our greatest worldwide dilemma is not declining natural energy supply nor pollution and a deteriorating environment, but an inadequate food supply associated with land as a limited and degrading resource.

Some countries already have well-documented information related to land use, which includes reports and maps on climate, geology, topography, cadastral survey, soils, vegetation, and agricultural, and forest practices. In contrast, however, many developing countries have relatively little information on which to base land use management and land use planning; and under these circumstances a phytogeomorphic basis to field survey and land management may be highly useful. The possibility of multiple land uses may not have been considered and probably the consequences of changes in land use will not be adequately understood.

In approaching the problem of land management and planning its potential use, it is necessary to survey the current land use and carry out an inventory of the land and its resources. In developing the land resources to their ultimate productivity, it is important that the form of land use and the management systems of different types of land are such that their level of productivity can be maintained indefinitely. In determining the most appropriate land use, it is also important to understand the dynamics of the system so that probable reactions to planned conditions can be correctly gauged.

For a given area of land we need to be aware of the current use and its potential use to support in the future a range of activities including changes in its general use (e.g., rainfed agriculture to irrigated crops; forestry to agriculture or vice versa) and its suitability to grow specific crops. In recognizing the need for action on this problem, FAO has compiled an agro-ecological zone map of Africa (Figure 2.3) for which unique soil–climate units were developed by overlaying climatic data on the FAO *Soil Map of the World* (Higgins and Kassam, 1981).

Unfortunately too little attention has been given to evaluating land potential. Land use mismanagement at the national and regional levels, which results in loss of soil fertility, soil erosion, crop failures, and associated economic problems, could often be avoided by assessing in advance the land potential in terms of its macroclimate, geomorphology, natural and induced vegetation types, and soil characteristics. In many tropical countries, for example, rainforest and high forest are being converted to agriculture without in-depth consideration of the future economic and ecological consequences. Often the conversion of natural forest to agricultural or savanna woodland to rangeland is uncontrolled or based simply on shrewd local observations of needs resulting from population increase, accessibility by road to the area, and ownership criteria.

12.2 MULTIPLE LAND USE

Multiple use occurs both in space and time. In regions of the world where rural development is very recent or when there is a long historical tradition possibly dating from biblical times, there is frequently resistance to viewing land as having multiple uses and to assume that the existing land use is the most efficient or economic. Most lands, however, are basically multipurpose, whether monoculture is practiced seasonally or over longer periods of time, or the same land is simultaneously used for more than one purpose, as in the case of agroforestry and in the Far East for rice cultivation and fish farming.

In the Mediterranean land zone, an example of multiple use is provided by the widespread seasonal migration (transhumance) between summer pastures in the uplands, now also used for recreation, and winter grazing in the lowlands after cereal and legume crops have been harvested. More than one crop is sometimes grown simultaneously on the same land, as in many tropical countries, and as on the Sorrento Peninsula in Italy, where there is locally a three-story culture on small terraces of fruit trees, vines, and vegetables.

Moreover, the same area may experience quite different types of land use in different historical periods. The central Tunisian steppes supported an essentially sedentary pattern of settlement based on cereal and olive cultivation in the Phoenician and Roman periods, changed to seminomadic grazing after the advent of Islam, and have been returning to settled agriculture as a result of increasing population and the adoption of European farming methods since the midnineteenth century. In the Mediterranean region it is often no longer economical to cultivate small terraces for hand-harvested cereals and even traditional olive groves in mountainous areas are being abandoned. The only solution may be reafforestation after 1000 years or more of agriculture.

12.3 GEOMORPHOLOGY

There are a number of ways in which geomorphic data can assist in land use and land evaluation studies. The effects of geomorphology expressed through its three fundamental components are rock materials, topographic form, and the processes that result from the interaction of these with climate. The general control that these exert on vegetal cover has been considered earlier, but the specific effects on each on current and potential land uses are also important.

First, different rock materials at the earth's surface may support different types of vegetation, which in turn reflect different land suitabilities. In the San Luis Obispo area of California, for instance, *Arctostaphylos* and *Ceanothus* spp. are selective of certain types of substratum and *Cupressus sargentii* is confined to serpentine outcrops (Cannon, 1960). Glacial deposits occur beside alluvium in the Weser Valley (Germany), but they visibly differ in the plant populations they support (Mensching, 1950). More generally, in unglaciated areas of southern England and northern France with their maritime climate, there is a clear distinction between the calciphilous vegetation of chalk lands and the conifers and heaths on sand deposits. The first are usually best suited to extensive cereal cultivation and the second to forestry and recreation. Under the semiarid conditions of the Middle East, by contrast, the best land for rain-fed agriculture often lies on a alluvial plains and rich volcanic soils, whereas limestones under the same climate have soils that are too shallow and droughty for cultivation and must be left to rough grazing.

Second, topography is a prime determinant of land use. Slope, altitude, aspect, and exposure influence the distribution of surface and groundwater, soils, and natural and induced vegetation. In upland Britain, for instance, there is often a descending toposequence from the hilltop with lithosols sustaining only mosses and lichens, open peaty moorland used for rough sheep grazing and grouse shooting, podzolized soils with conifers on the hillslopes, brown forest soils with mixed farming and deciduous woodland on the piedmonts, and wet meadowland with gleyed soils in the valley bottoms. These categories reflect land suitabilities with progressive decrease of values away from the well-drained lowland areas.

Similarly, toposequences in the tropics can be quoted to illustrate the effect of topography on land use. In Tanzania between the Usambara Mountains and the neighboring Pangani River Valley that flows along their western foot, the well-watered but hilly plateau top develops oxisols (red tropical soils) that support a varied tropical agriculture of small farms and large commercial plantations. The upland is edged by a steep mountain slope, bare except for precarious shrubland, which ends abruptly in an alluvial apron where a more restricted range of drought-tolerant crops including maize, millets, and sisal can be grown. Thereafter the Pangani River meanders along the foot of this apron across a clay plain rendered presently unsuitable for agriculture because of swamp conditions. The plateau top and alluvial lowlands are relatively valuable land, although the former suffers from topographic and the latter from drought limitations. The mountain front cannot be used because of its steepness and the clay plain because of its wetness.

Slope also has a strong influence on the mechanization of activities. At a point where the slope becomes too steep to practice mechanized cultivation for agriculture, the land may have to be devoted to forestry, although the soil conditions may still be suited to farming. Where the slopes become precipitous, they may be too hazardous for mechanized timber production and can best be maintained as protection forests to water catchment and as a part of wildlife management.

Third, geomorphic processes can be categorized generally as beneficial or detrimental. They may also be categorized as erosive or depositional and they may be slow, persistent, or occurring suddenly and cataclasmically. All processes affect land use and land use potential and have implications for conservation and development. Alluvial deposition can lead to the accumulation of usable soils in lowlands. The redeposition of materials eroded from coasts may lead to the formation of new lands, which may be reclaimed for agriculture, as can be seen around the coasts of Bangladesh, Holland, and around the Wash in eastern England.

The general effect of detrimental processes is to remove materials from a useful place and redeposit them where they are not wanted. A capability for predicting such hazards is a vital basis for developing warning systems and the initiation of protective and reclamation methods. Geomorphic evidence of the upper limits of flash floods (e.g., bench terracing) may be critical to wise urban planning. Water erosion includes mass movements, sheet wash, and gullying with redeposition of sediments on lower lands, sometimes in arid areas accompanied by salinization. Wind erosion includes both deflation and dune formation. Lands near to the sea may be lost through coastal erosion, especially where soft rocks are exposed to strong onshore waves and currents. Flash floods in rivers or exceptionally high tides can cause widespread destruction of riparian or coastal property and strong winds may cause dust bowling in semiarid areas in the absence of well-established plant cover.

12.4 VEGETATION

Similarly, there are a number of ways in which vegetation data contribute to land-use studies and land evaluation. The vegetation sensitivity mirrors conditions of the natural environment as they currently exist. The natural vegetation is often an excellent expression of the total effect of environmental factors. The physiognomy of the plant community, its species composition, including the dominant species, and the seral state of the vegetal succession all contribute to this perspective.

The importance of observing changes in the physiognomy of the vegetation is frequently ignored as an integrating factor portraying the effect of the interaction of climate, soils, and landforms. In an identified climatic zone, it is frequently possible to appraise crudely the mean annual rainfall in terms of changes in plant physiognomy (e.g., life-forms, stand height). The abundance of lichens and mosses on the tree and the presence in the tropics of tree orchids may signify a climatic mist belt. Similarly, changes in the dominant species may be indicative of other climatic factors important in land use such as local frost hollows and exposure as influenced by coastal winds. In fact, the vegetation can be viewed as a barometer of the interaction of environmental factors averaged over a period of time.

Also, as mentioned earlier, the vegetation frequently reflects the soil characteristics, and hence this recognized correlation is used advantageously by the soil surveyor to extrapolate horizontally information on the soil profile. Provided a threshold of information is already available on the performance of vegetation according to soil types, the vegetation can frequently be used directly to provide quickly the areal information on the soil characteristics. This may cover changes in soil types, soil depth, and the presence or absence of a hardpan or a high water table. Soil surveyors complete their soil map by noting in the field the vegetal changes between line transects or point samples and avoid the somewhat tedious task of digging additional soil pits or making additional auger drillings.

In land use studies, emphasis is usually placed on the physiognomy of the vegetation and not on the floristics of the plant community, although the importance of indicator species must also be taken into account. Thus changes in the physiognomy and the dominant species of plant communities on a flood plain can indicate for land use planning the limits of seasonal flooding, flooding that occurs every few years and possibly the extent of flash floods occurring only a few times in a century. Within a land unit the study of vegetal succession and the identification of the remnant natural vegetation occurring with the induced vegetation of current land use can be used to help map the areal extent of the potential land use and the site quality.

12.5 PHYTOGEOMORPHOLOGY

If we map the landforms and vegetation separately and then proceed to integrate these two factors from the separate maps, this may be difficult without recourse to further fieldwork, because it may be difficult or impossible to reconcile the differences in common boundaries shown in the two thematic maps. A similar problem occurs when an attempt is made subsequently to reconcile the mapped data of a soil survey with mapped data of landforms or vegetation. Hence it is important to record in the field landform, vegetation, and other data simultaneously.

We have already observed that the phytogeomorphic factor provides a sound basis to the integrated approach to dividing the landscape into meaningful units, which have a uniformity of geology, geomorphology (including topography), vegetation, climate (macro and micro), and soils. We have also observed that in mapping the landscape, meaningful land units and boundaries imbuing the phytogeomorphic factor can be achieved by pursuing an integrated approach.

When the landscape has been mapped simultaneously in terms of its phytogeomorphic characteristics and is later appraised in relation to current land use and land potential, it will be found that (a) at each level of subdivisive hierarchical classification, meaningful land units have been identified and delineated; (b) according to the intensity of the land use planning, the level of the required land unit can be identified and the hierarchical classification cut off at that level or pursued later in more detail to provide smaller land units; (c) the larger land units (e.g., land province, land subprovince, land system) can usually be closely associated with current land use and the smaller units (e.g., land catena, land facet, and land element) with the potential land use, (d) land potential of extensive areas can often be assessed from the phytogeomorphic units without recourse to detailed objectives of management, provided their general use is identified (e.g., forestry versus agricultural crops versus grazing); and (e) land potential of local sites cannot be assessed without carefully identifying their use. This usually requires identification of specific objectives of management so that the smaller land units can be grouped into meaningful management units (e.g., Japanese larch on bracken slopes in west Wales).

12.6 SYNTHESIS OF LAND UNITS

The grouping of subdivisive land units into larger units will often be necessary for land use management and land-use planning, particularly when economic and political constraints are applied. Frequently, the same crops will be observed to be grown in several different land systems and hence for land-use planning it is necessary to group the simple land systems together into units of equal site quality for the specific crop. If economic conditions change or new crops are introduced, land use planning may require further studies of the individual simple land systems and their regrouping.

Assisted by computer-programmed synthesis, phytogeomorphic data sets can now be analyzed and regrouped into larger units according to the objectives of the land management. Several statistical techniques are used. Two principal approaches will be described. Other statistical techniques are provided by Sneath and Sokal (1973). Once a synthesis scheme has been developed for a statistical population, new sets of data descriptive of land use can be assigned objectively using discriminant analysis (Morrison, 1976).

First, the principle of cluster analysis has been successfully applied to land classification (Driscoll, Bettors, and Parker, 1978). This statistically combines individual units into larger groups for identified objectives of management. The units are aggregated on the basis of their overall similarity as rated by indices such as correlation coefficients and euclidean distance (Severson and Thilenius, 1976). Results can then be presented hierarchically according to the level of the similarities. In a diagrammatic presentation the results will resemble a dichotomous key.

Second, the principle of ordination (Bray and Curtis, 1957) can be invoked for consolidating land units. This consists of plotting the phytogeomorphic data in the two, three, or multidimensions of a coordinate frame and having smaller dimensions than the original data. Unlike cluster analysis, ordination does not group sites into clusters of correlated data, but assumes the data are in continuum and hence form natural groupings. Ordination of land use data enables the original data sets to be grouped into a few principal components and enables an assessment to be made of the relationships of the land units.

For management and planning purposes the regrouping of land facets and land elements into larger management units is common with changes in local land use. This is particularly seen where government policy leads to new economic or social factors. Several examples will be given from the nineteenth century onward.

Frequently under rainfed agriculture the smallest unit of management will be the land facet, but when irrigation is introduced, the management unit is likely to be the grouping of land elements. In the United Kingdom the regrouping of upland facets has occurred with social and economic changes from upland grazing to coniferous plantations; and where, as in West Wales, the steep slope land facets have been traditionally under oak coppice, different grouping of the facets may be needed according to site quality when replanted with different coniferous species (e.g., Japanese larch, Douglas-fir, Norway spruce). Again, the demanding requirements of certain types of agriculture may place emphasis on the importance of the land element as the unit of management and not the land facet or group of land facets. In the Beaune–Dijon area of France, the best traditional wine-producing lands coincide with land elements on the intermediate slopes.

A consideration of Basilicata Province, Italy, comprising the southern end of the Apennines is relevant to the application of phytogeomorphology to land-use management and planning (Figure 12.1). Substantial changes in land use have been associated with changes in the occupational structure of the population of recent years. The land region is hilly and is underlain by much disturbed Jurassic and later sediments rising to heights of over 2000 m, from which a series of relatively parallel rivers, running from north to south, drain into the Gulf of Taranto.

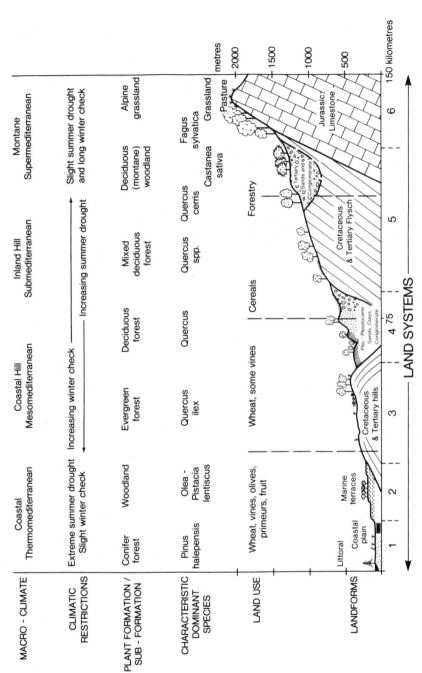

Figure 12.1. Schematic phytogeomorphic transect of Basilicata Province, Italy.

177

The land systems have traditionally supported a relatively intense agriculture based on wheat, vine, and olive on the lower land facets, more exclusive wheat cultivation at intermediate altitudes, pastoralism and forest on the steeper and higher altitude land facets, and limited rough grazing on alpine land facets of the summits. The trend since World War II has been twofold. First, land use intensities have been polarized. There has been an industrialization of agriculture with intensified fruit and primeur cultivation aided by irrigation in the valley land facets and land elements, coalescence of holdings and mechanization of cultivation along the coast, on the marine terraces, and at lower altitudes on the hills; and a decline of agriculture with reversion to forest on the upland land facets accompanying the most acute rural depopulation.

Second, the area has become an important source of water not only for local irrigation but even more for exportation to Apulia through a system of dams and barrages involving all five rivers. This has led to an emphasis on catchment management to control erosion and to maximize and to even the flow over the year, thus focusing attention on such measures as afforestation and gully control.

Land use classifications have close relevance to these complex changes. The intensification of lowland agriculture increases the need to survey the land facets and land elements and to group them into *management units*. The changing type and intensity of land uses with rising altitudes and steepening gradients demand an emphasis on grouping land facets in ways relevant to forestry and hydrology, except where it is important to discriminate slope and lithological features that affect runoff and infiltration, and soil depth and climatic factors that affect arboreal productivity. It may be found convenient to identify the purpose of such groupings by prefixing the word management to the land unit concerned. Thus we have *management land facets, management land catenas,* and *management land systems*.

THIRTEEN

APPLICATIONS OF PHYTOGEOMORPHOLOGY

13.1 INTRODUCTION

Phytogeomorphology has been considered briefly in the general context of land-use management and planning. However, there are more specific discipline-oriented studies that can benefit directly or indirectly from a phytogeomorphic approach. Some of these will now be considered.

It will be observed, as mentioned previously, that the balance between the importance of landforms and vegetation varies greatly. In some cases vegetation is the dominant factor, whereas in others the study of landforms provides most of the information. It is valuable to involve both landforms and vegetation simultaneously in specific field problems without prior assumptions about their relative importance. In this way the maximum use can be made of phytogeomorphology. As stated earlier, workers can pursue a hierarchical approach by subdividing the landscape into integrated land units or they may examine landforms and vegetation in parallel.

The formal application of an integrated concept of geomorphology and vegetation to land-use planning is relatively new and few published examples of the practice can be quoted. In practice the integration has often been used intuitively to make planning decisions, but will seldom have been recorded stepwise during the process.

This chapter gives a number of examples of the application of phytogeomorphology to different practical disciplines. The list cannot be comprehensive and varies both in the type and amount of phytogeomorphologic input between the different disciplines. For some disciplines, such as geology, soils, and engineering, landscape units are themselves the direct basis for planning. In others, such as hydrology and mineral exploration, they mainly serve as indicators of the collateral environmental information that guides the seeker to the locations where the resources may be found or controlled. In all cases, however, they provide a starting point for investigation, survey, and planning.

A sound grasp of the scientific basis of phytogeomorphology may prove useful to those concerned with economic and social theories and practices. There has, however, been no attempt to introduce these aspects separately into the text, although there is need for their consideration by those concerned with the application of phytogeomorphology to everyday problems. Conversely, it is recommended that whenever possible, in applying photogeomorphology to the planning and management of the natural landscape, due consideration be also given to the economic and social factors and their impact.

13.2 CLIMATE

Differences of opinion have existed since the late nineteenth century on whether the classification of climate should be based exclusively on climatic criteria or take vegetation into consideration. Also, the effect of climate is influenced by the landforms. At least for those concerned with land-use planning, the combination of climate, landforms, and vegetation is realistic and, as mentioned earlier, vegetation data have often been used to improve relatively scanty climatic data. The close relationship between climate and vegetation is also seen in the literature in the use of such expressions as savanna climate, tundra climate, mist belt forest, and alpine meadows.

The advantage of associating vegetation with macroclimatic mapping is that it helps overcome the problem of selecting the most appropriate climatic criteria and may help in supplementing data from sparsely distributed climatic and meteorological stations. Certain physiognomy of the vegetation or one or more species of the plant community can be used to help identify climatic factors and to extend climatic boundaries identified on the basis of climatic field stations. Particularly as the climatic mapping scale increases, attention also needs to be given to the effect of the landforms, including micro- and macrorelief on the microclimate.

The physiognomy of the vegetation and its species composition are the product of a complex interaction of environmental factors over a period of time far exceeding that of written climatic records, which unfortunately are usually limited to daily temperature changes, daily precipitation, and occasionally relative humidity. Also some meteorological stations (e.g., at airports) may not be representative of the agroclimatic zone with which they are associated through their ground location.

The trained plant ecologist is able to use effectively the varying expressions of the vegetation as a barometer of climatic changes. The reading of changes in the vegetation as related to climate is most readily seen in tropical mountainous areas with changes in altitudes and major changes in aspect. Usually the physiognomy of the vegetation is conspicuously different on north and south slopes at higher latitudes and on leeward- and seaward-facing slopes. Probably the best estimate of exposure to the wind is achieved by studying vegetal physiognomy. Even where the differences are subtle, minor differences in the height of the mature vegetation and the growth rate of the dominant woody vegetation may be helpful indicators of climatic variations and often presence or absence of dominant species or their condition can be used to indicate frost hollows and other limiting climatic factors.

13.3 GEOLOGY

Phytogeomorphic criteria provide data for geological exploration and survey, not only for the normal mapping of stratigraphic units but also for that of mineral and water resources. The geological literature contains many examples of the ways in which surface form and vegetation can be used to indicate geological structure, materials, and stratigraphic relations. The presence or absence and the changes in the physiognomy of the vegetation can indicate geological features such as bedding, dip, foliation, folds, and faults. Although materials can vary widely so that the same rock type assumes very different surface forms in different areas and may even form a hill in one place and a valley in another, there are nevertheless certain guidelines that generally hold. Thus in areas of sedimentary rocks, higher ground indicates competent materials such as chalk, limestone, sandstone, and quartzite, whereas valleys reflect the presence of the more easily eroded shales, mudstones, slates, and micaschists. Where sedimentary and metamorphic rocks occur together, they may be separately identifiable because the lineaments controlled by bedding tend to be longer, less numerous, and more evenly spaced than those controlled by foliation.

The eastern coastal plain of the United States illustrates readily the way in which landscape indicates underlying geology. In New England the higher ground tends to indicate the presence of Precambrian gneiss and granite, with the latter forming many of the isolated mountains. Moderate altitudes are associated with Paleozoic metamorphic rocks, low ground with unmetamorphosed Paleozoic sediments, and basin sites with structural troughs in Triassic rocks. These topographic differences are further emphasized by an altitudinal zonation of the vegetation types.

Southward from the Cape Cod area the landscape is arranged in belts of cuestas and valleys roughly parallel to the inner and outer edges of the coastal plain. These indicate an outcropping sequence of geological deposits younger than the Triassic dipping gently seawards. Four belts from west to east across New Jersey indicate the geological sequence. An inland clay lowland, intensively cultivated for vegetables and adjoining a well-drained sandy belt, identifies the Cretaceous strata. These are followed by an intermediate belt on Tertiary and a coastal one on Quaternary deposits. The glacial origin of the latter is discernible from its extension into the fishtail ends of the Long Island coast, the northern continuing through Cape Cod and the southern through Martha's Vineyard and Nantucket. Their uneven topography and the absence of intensive agriculture helps identify them as moraines composed largely of sand and gravel deposits.

Close relationships between landform, land use, and underlying geology can also be seen in the English Midlands where a series of northeast–southwest trending cuestas and strike vales has distinctive land uses and land capabilities that indicate a sequence of beds of limestone, sandstone, chalk, and clay dipping uniformly and generally falling in altitude toward the southeast. The Cotswold Hills whose summits, which do not exceed 400 m, are used for arable farming and whose western escarpment overlooks the Vale of Severn identify the outcrop of massive Oolitic limestones. These are succeeded first by the lower and more fertile arable lands on the Cornbrash outcrop and then by the wet pastoral lowland of the Vale of Upper Thames that indicates the presence of the Oxford Clay. Southeast of this is a low upland cut transversely by the Thames at Oxford, whose mixed land uses range from market gardening to golf courses, with these being associated with the soft sandstones and mixed farming with the rubbly limestones which alternate but together form the Corallian outcrop. This cuesta dips in turn into the wet pastoral Vale of White Horse on the Kimmeridge and Gault Clays, then rises first to the fertile fruit-growing bench of the Upper Greensand, and then to the open treeless landscapes with little surface drainage and large arable fields of the Chiltern Hills and Berkshire Downs which are coterminous with the chalk outcrop.

13.4   MINERAL EXPLORATION

Mineral resources tend to be related to the landscape in predictable ways. They are occasionally visible on the surface, but are more usually interpreted from an understanding of the geology, which may be reflected in the landforms or vegetation. Most phytogeomorphic work has hitherto concentrated either on the landforms or the vegetation and there has been relatively little integration of the two.

At regional scale ore deposits are related to rock outcrops, faults, and shear zones. This may be associated with particular phytogeomorphic changes and the directional changes in surface lineaments recorded on remotely sensed imagery.

At a more locale scale, sources of construction materials, such as granite or limestone outcrops and unconsolidated sand and gravel deposits, have recognizable surface characteristics. These are sometimes associated with particular landforms and vegetation types, including various forms of stunting or chlorosis of vegetation resulting from the effects of the underlying rocks.

Higher value minerals, which usually lie deeper below the surface, can be interpreted from landforms that indicate the underlying stratigraphy. Coal is a sediment whose beds can be traced from the direction and nature of their outcrops. Oil is located under sedimentary folds and domes of which topographic indications are often visible at the surface. Other economic minerals are related to specific points in the landscape. Relatively immobile residual concentrations of aluminium and nickel may sometimes be found in landforms from which other materials have been removed. Valley terraces and stream beds may have placer deposits of gold, platinum, and tin, and basin or lacustrine sites may have concentrations of iron, manganese, and occasionally base metals and uranium selectively precipitated from solution.

Geobotany has proved a useful science in mineral exploration. It uses vegetation as an indicator of rock type either by changes in the species distribution, in the physiognomy of a stand, in the color of foliage of individual plants, or in the mineral content of the leaves. Furthermore the accumulation of minerals in plant tissues is followed by their concentration in the soil, according to the *Goldschmidt enrichment principle* (Huang, 1962). This states that during the normal process of supergene enrichment in forested areas, minerals dissolved in groundwater are absorbed by roots and deposited in plant leaves as they transpire. When these leaves fall and decay, the less soluble mineral contents are concentrated in the humus layer by the differential removal of the more soluble mineral contents in drainage water, and constitute a zone of surface enrichment. Spectrographic analysis of this layer has led to the location of tin and tungsten deposits in the British Isles and of chromium in Greece.

Plant and tree distributions and 'indicator flora' may locate rock types favorable to ore deposition. Limestones have notably calcicolous vegetation. Some metalliferous rocks strongly enrich the plants that grow on them. As extreme examples, some plant species can absorb up to 0.7% lead, others up to 1% zinc, and still others up to 3.2% manganese from their soils. No less than 114 plant species, mainly among those endemic to serpentine rocks, have been shown to be able to accumulate nickel to concentrations of over 1%. Analyses of the nickel content of *Nothofagus fusca* plants, for instance, in Nelson Province, New Zealand, showed close correlation with its level in underlying soils and thus in the rocks from which these were derived (Brooks, 1983). Some plants are poisoned by high concentrations of metals such as lead, nickel, and copper so that they are either absent from stands in which they would normally be expected or they appear in damaged condition. Excess boron can, for instance, cause deformation or stunting, bitumen can lead to gigantism, and elements such as chromium, cobalt, copper, manganese, molybdenum, nickel, and zinc can give rise to distinguishable types of leaf chlorosis. The successful location of the vast copper deposits in Haute Katanga (Zaire) in the 1940s resulted from the observation on stereoscopic pairs of aerial photographs of stunted vegetation and species distribution in the field that could not be associated with changes in soils, climate, or geology. In Western Australia the location of alumina deposits has been observed on aerial photographs to be associated with dry sclerophyllous forest dominated by *Eucalyptus marginata*.

Vegetation–rock associations indicating minerals can also be seen on remotely sensed satellite imagery. Ballew (1975) used multispectral digital Landsat data to identify relations between both vegetation cover and bare rock on the one hand and geochemical data on mercury, lead, silver, copper, and bismuth on the other in the Washington Hill mercury mining district. Metal anomalies corresponding to alteration patterns around known prospects were identifiable from the greater relative frequency of ponderosa pines (*Pinus ponderosa*), which was often associated with soils of low pH and low available phosphate. Similarly, using the biomass index obtained from the MSS7–MSS5 ratio on Landsat imagery, of Quebec Province, Canada, it was found possible to identify hills composed of serpentine associated with asbestos and chromium ores because of the chemical unfavorability of their soils to plant growth (Brooks, 1983). Ores associated with a high content of phosphate are sometimes recognizable from its effect on plant growth. For example, the contrast between the luxuriant vegetation on the phosphate-rich carbonatite deposits at Sokli, Finland and that on the nutrient-poor skeletal soils derived from the surrounding Archean basement rocks led to their separate identification (Paarma, Vertiainen, and Penninkilampi, 1977). Similarly, Brooks (1983) has summarized research that shows that, in semiarid areas, kimberlite can be identified from the richer plant growth associated with high soil phosphate and potassium.

13.5   ARCHEOLOGY

The importance of phytogeomorphology to archeology lies in its use for recognizing and categorizing land types and improvement patterns as a means of identifying and reconstructing the past. The two distinct scales at which this can be considered are regional cultural units and local settings for families and small groups. Frequently the cultural units of the past can be observed on aerial photographs as being associated with recognizable land systems, land catenas, and land facets.

At regional scale, groups of people have activity loci within districts somewhat analogous to the catchment areas of rivers within which movement, settlement, development, and interchange take place relatively rapidly. These activity loci are often lake or river basins or ancient plains that coincide with phytogeomorphic regions, land systems, or specific land facets. The settled areas may have been eroded or buried as a result of tectonic or volcanic activity, the effects of which are especially conspicuous in coastal areas where there has been emergence or submergence.

The potentially large amount of archeological data about many regions necessitates the use of a sampling program either to select areas representing a microcosm of the region as a whole, or purposive sampling of selected land facets, or else to sample a particular activity system such as agriculture, the crafts, trade, and communications wherever it occurs. In each type the sampling patterns are stratified in terms of relevant landforms, plant communities, and water resources.

Within regions the individual sites may be visible or invisible at the surface, or visible on aerial photographs through expression in the vegetal structures. In recent years there has been a rapid increase in the number of known sites, largely as a result of their expression in the form of vegetation patterns or occasionally species distributions as observed on aerial photographs. To find these unknown sites, one starts from the assumption that as the surface indications of most ancient land uses resemble those of similar land uses today, their traces can usually be interpreted as related to recognizably homologous land units. Road layouts, village settlements, seafaring facilities, and military constructions, for instance, are often associated with particular topographic conditions, and patterns of settlement and agriculture reflect the locations of fertile soils, water sources, vegetation types, and ore-bearing rocks.

Specific aspects of both geomorphology and vegetation can be used to identify archeological sites, especially when viewed on aerial photographs. The landforms, as expressing the limitations of land systems and specific land facets, can be associated with archeological sites, whereas the vegetation not only confirms these observations but may also provide an outline of the patterns of settlement or of individual buildings when studied on aerial photographs. For example, small banks and terraces from ancient field patterns show up especially well with oblique lighting. Hill forts in northwestern Europe are sometimes identifiable from tree clumps as at Chanctonbury, Sussex, and on the Sinodun Hills in Oxfordshire. Aboriginal shell middens indicating gathering points are recognizable on large-scale photographs taken in Tasmania.

Information given by topographic features is increased where it is also possible to discern differences in plant growth. Abandoned houses and settlements often show a luxuriant growth of weeds or different plant species on former kitchen middens, although this effect diminishes relatively rapidly within an archeological time scale. In general, ancient excavations such as pits, trenches, and post holes deepen the soil and make for better growth, whereas buried foundations, floors, and road metaling have the opposite effect. The resulting contrasts are most marked in the dry season when soil moisture tension is at a maximum and significant archeological discoveries have been made at times of exceptional drought. In dry country especially, disturbed ground may carry more moisture. This, for instance, led to the discovery of Ptolemaic irrigation channels in the Fayum (Egypt) by noting the growth of desert plants mainly of the genus *Mysembrianthemum* (Clarke, 1964). The agricultural crops most sensitive to these effects are often cereals or short-term leys. In cool temperate latitudes the best time to see such differences is during the latter half of the growing season when crops have well-extended roots and are growing vigorously, or when light snow on the ground is melting, or in periods of excessive moisture or moisture shortage. New Roman and Norman sites have been recorded on aerial photographs after prolonged droughts in England and southern Scotland.

13.6 SOIL SURVEY

Because soils are natural bodies that closely reflect their climatic, geomorphic, and vegetal environments, phytogeomorphic interpretation is a fundamental survey tool both in their recognition and in determining their suitability for different land uses. The role of landforms and vegetation as associated with vegetation has been considered in Chapters 3, 4, and 12.

The landforms and vegetal cover exert some control over soil formation. Natural vegetation affects soils, and differences in the nature and amount of organic matter are especially marked across major ecological boundaries such as those between forest and grassland and between cultivated and uncultivated land.

The landforms affect soil formation through the influence of slope on processes such as solifluction and colluviation by which soils are thinned on upper slopes and deposited on lower slopes, and by a general increase in groundwater and drainage impedance on downslopes. Such toposequences have long been recognized in soil surveys in the use of *soil catenas* in the tropics (Milne, 1935) and *soil associations* as mapping units in areas subject to hydromorphic conditions such as Scotland (Glentworth and Dion, 1949). The rock of which the landform is composed also provides the parent material of soils and even in mature profiles often governs their fertility through controlling, in combination with climate and vegetation, their depth, texture, and base status.

The impact of phytogeomorphic conditions on soil formation can be seen at a variety of scales. Detailed soil surveys at the national and local levels in many countries show close correlations between soils, terrain conditions, and potential land uses. The Sudan, for example, presents an environment where clear relationships exist between four major land resource factors: rainfall, geomorphic site types, soil textures, and tree species. In fact, when three of these four are known, not only can wide-ranging inferences and predictions be made about the fourth factor, but distributional anomalies can be identified that lead to further knowledge.

The Sudan is a large intertropical country situated almost entirely on the relatively level African Shield. Rainfall decreases regularly northward from more than 1400 mm in the extreme southwest to less than 25 mm in the far north, a progression closely reflected in a regular zonation of the natural vegetation from broad-leaved woodland through progressively sparse savanna and steppe to barren desert. Aridity is such an overriding ecological limitation that a study of the internal variations within the country presents opportunities for identifying and grading the site and soil factors influencing plant growth. The natural tree species provide a good indicator of these conditions.

In the northern two thirds of the country, vegetation-site relationships give such a clear indication of rainfall and soils that they can be used for predictions. Hills and valleys give soils that differ from the local datum, especially where the moisture-concentrating effects of a low site are combined with the higher infiltration capacity of stony or sandy materials. *Acacia senegal* occurs on sites ranging from hillsides in Kordofan and Rufaa (15°N, 33°E) where the rainfall is 300 mm to sheet erosion slopes at Nimule (4°N, 32°E). *Acacia fistula* ranges from plains in Kassala province with 500 mm of rainfall to swamps in the Upper Nile province with 1000 mm, and *Acacia tortilis* is found in seasonal runnels in the desert with a rainfall of 50 mm, on datum sands at Khartoum with 150 mm, on clays in Kassala province with 300 mm, and on erosion slopes in Butana (14–16°N, 34–36°E) with 500 mm. In short, it was found that with no recorded exceptions, all species progressed through their rainfall span via the same sequence of site types, and given the same rainfall, the site types fell into the same sequence of moisture favorability for tree growth (Figure 13.1; Smith, 1949).

The general validity of these rules has led to the discovery of an anomaly. Within the clay plains (site type F), the sequential change of tree species across the isohyets between Khartoum and the Ethiopian border is somewhat less marked than expected. This is because the southward increase in rainfall is more rapid than is that of the soil water available to plant roots. The explanation is that the percentage of clay in the soil increases with the rainfall and in-

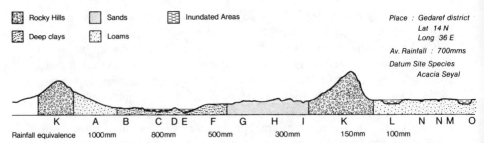

Figure 13.1. Sudan: the site transect that shows the favorability of sites throughout the country for tree growth (after Smith, 1949). All species progress through their rainfall span via the same series of site types and in the same order of site types. The diagram shows the series from A to O that is described in the table below. The order and the rainfall span within these sites can be equiconditional, as evidenced by the growth of a single species on all of them. Comparison of the site type with the rainfall shown below it gives a measure of the comparative values, in terms of moisture availability, of the various site types.

	Site Type Description	Typical Tree Species
A	Hard-surface slopes, i.e., sheet slopes	*Acacia mellifera, A.*
B	High, old flood plain; if now flooded, only for days	*senegal, Balanites aegyptiaca, Zizyphus spina-christi.*
C	Low flood plain, flooded for weeks at a time	*Acacia fistula*
D	Mounds, usually of termite origin, in swamp	*Bauhinia reticulata*
E	*Mayaas* and *rahads*, i.e., land-locked pools	*Acacia arabica*
F	Clay plains, no runoff and no standing water	*Acacia seyal*
G	Mature sand plains, no runoff and no standing water	*Sclerocarya birrea*
H	Immature sand, new dunes, and old sand-hills	*Albizzia zygia, Detarium*, sp.
I	Small well-watered and well-drained pockets in sands	Not represented in the *Acacia seyal* transect
K	Hills of rough, rocky surface	*Boswellia papyfirera, Lonchocarpus*, sp., *Sterculia*, sp.
L	Wadis or large seasonal watercourses flushing for an hour or two after rain	*Acacia sieberiana*
M	Hard plains of grit or rock	Not represented in the *Acacia seyal* transect
N	Seasonal runnels flushing for an hour or two after rain	*Pseudo-cedrela kotschyii*
O	Banks of permanent streams or rivers	*Acacia campylacantha, Ficus sycomorus, Tamarindus indica*

deed has an approximately constant relation to it on datum sites. Subsequent research in the northern Gezira indicated an increase in soil clay percentages southward from the neighborhood of Khartoum (Hunting Technical Services, 1963). These trends run opposite to the expected direction of fining outwards from the source regions south of Roseires and in Ethiopia. Thus a study of the relationship between geomorphic site types and vegetation leads to the significant conclusion that clay minerals, notably montmorillonite, are being synthesized in the soil to a degree directly related to the rainfall.

13.7 SOIL DEGRADATION ASSESSMENT

Water erosion is the chief degradation process in all areas of high or medium relief and also in those areas of low relief where there is vigorous fluvial activity such as is evidenced by incised valleys, undercut banks, or active alluvial or colluvial deposition. Wind action is dominant wherever wind scour, deflation, and dunes are more important than water erosion. Salt accumulation is dominant in arid climates where dischargeless basins are inundated from contributing catchments that leave surface clay and salt pans on evaporation.

The FAO-UNEP-UNESCO project for worldwide soil degradation mapping, commenced in 1975, covered the occurrence of these forms of degradation in parts of North Africa and the Middle East at a scale of 1:5,000,000 and included experimental maps and assessments at a scale of 1:1,000,000, based on a phytogeomorphic interpretation of Landsat imagery which emphasized features related to soil degradation. The approach used in interpreting the remotely sensed imagery involved the recognition of both landforms and vegetal cover.

The method adopted at 1:5,000,000 was to map a series of sample areas covering about one fourth of North Africa and the Middle East, developing legends for morphodynamic units and vegetal cover and land-use units. For water and wind erosion, it involved an emphasis on soil factors such as texture, permeability, and organic matter, vegetal cover, and land use, and in addition for water erosion those of slope gradient and slope length. The final morphodynamic legend distinguished between landforms dominantly experiencing erosion and deposition and subdivided these on the basis of form, lithology, and geomorphic processes associated with soil degradation.

The vegetal cover and land-use legend recognized that the risk of soil degradation increased as vegetal cover decreased and as land use intensified, both of which are generally recognizable in the landforms portrayed on remotely sensed imagery. The mapping units were a combination of percentage vegetal cover with type and intensity of land use. Both morphodynamic and vegetal cover and land-use legends were tested for accuracy and comprehensiveness in the mapping of Iran (Mitchell and Howard, 1978).

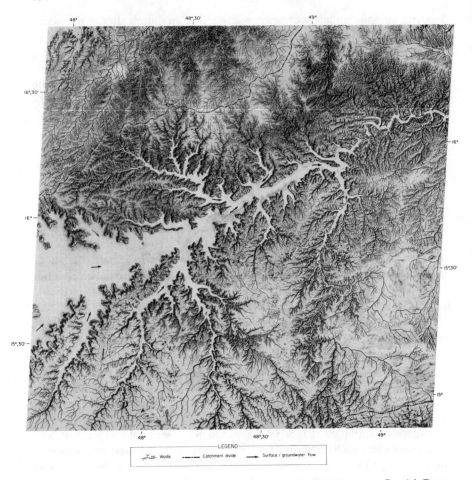

Figure 13.2. Interpretation of the drainage network of the Wadi Hadhramaut, People's Democratic Republic of Yemen (Travaglia and Mitchell, 1982).

13.8   WATER RESOURCES

Phytogeomorphic data can help to indicate the presence or location of water resources and may contribute to its assessment. It is often possible to infer the location of water from the nature and orientation of phytogeomorphic units. The straightforward approach is by the quantitative analysis of the way in which rainfall and snowfall on the particular landform assemblage within the catchment are distributed into infiltration, runoff, and evapotranspiration.

A FAO pilot study of the western part of the People's Democratic Republic of Yemen indicates the usefulness of phytogeomorphology in studying water resources under arid conditions (Travaglia and Mitchell, 1982). Landsat imagery was used to provide a drainage pattern analysis of the Wadi Hadhramaut area and a lineament analysis of the western part of the country.

A study of the intricate drainage network of the Wadi Hadhramaut tributaries and the exact location of the watersheds enabled an assessment to be made of the likely distribution of the water (Figure 13.2). Elsewhere, under the sparse rainfall of the region, vegetation thrives only where groundwater is available and this reflects geological lineaments. Open fractures appear to act as groundwater conduits, carrying the somewhat higher rainfall falling on high ground out onto the plains and sustaining a thin growth of perennial vegetation along their courses. The recognition of such features permitted the identification of four areas, justifying future field investigation in the central part of the country south of the Ramlat Sabatayn dune field.

13.9   FLOOD PLAIN MAPPING AND MANAGEMENT

In the mapping and management of flood plains, phytogeomorphology combined with remote sensing can provide an important base in identifying land units according to their flooding liability. Water portation, urban development, and farming methods are all influenced by the landforms of flood plains.

Flood plains occur in areas adjoining stream and river channels. As streams mature in geological time, the valleys widen, watersheds enlarge, and drainage increases. Again, due to the population's activity, runoff may increase, for example, after clearing of the forest and according to the farming practice. As a result, and particularly during pluvial periods, there is an imbalance and flooding occurs. Young streams may be fast-flowing and stabilized to one channel; old-age streams may persist with one channel but meander and change course. Mature streams have intermediate speeds of flow and widths of meander plain. All will be liable to flood following periods of heavy rain or snow melt. Then, the rate of erosion may accelerate and the rivers carry larger amounts of suspended material that are deposited on the flood plains and thus increase the elevation of the river system. In times of abnormal runoff, extensive areas will be inundated and when inundation is fairly frequent, the flood plain will develop distinctive natural vegetation that is tolerant of flooding.

Three important geomorphic units, each associated with distinct plant communities, are associated with flood plains. The first is the *natural levee* formed on the banks of streams during periods of overflow; the second is a level area (*cover flood plain*) seasonally covered by low velocity flood waters and marked by scars left by currents at the time of flooding; and the third are *slack water deposits* consisting of slight depressions and obvious wet soil conditions (e.g., oxbow lakes at various stages of fill) along valley walls, at the base of terraces, or at the margin of higher elements of the flood plain. With the aid of vegetation, whether natural or induced, it may be possible to divide these further into three categories according to the periodicity of flooding. The first category occurs annually when flooding is minor; the second category occurs less frequently, perhaps every two to five years, and the third at times of flash floods, perhaps every 10 to 20 years or even less frequently.

Depending on population pressures and the intensity of agriculture, the flood plain may have been cleared of its natural vegetation, but frequently the crop types and land management can be used to separate annually flooded areas from less frequently flooded areas. There is, however, a danger that urban development will fail to recognize the danger of occasional flash floods that may only occur a few times or even once in a century. Where the natural vegetation persists or remains undisturbed, the plant communities on flood plains are usually distinctive both in their physiognomy and floristics. There is often a dense tree or shrub cover on wet sites that include seasonal swamps. Areas flooded annually and exposed to strong currents will usually be dominated by ephemeral species and woody species may be sparse or absent. Where large trees occur, flood scars may be seen on the tree boles. On slack water deposits, the in-fill of the oxbow lakes may provide a complete sere from water plants to closed forest.

13.10   FORESTRY

Forestry involves both forest inventory and improving the management of the forest. In this respect, the classification of the existing forest, or the land for afforestation according to its site quality, is important. Forest site qualities can usually be recognized by the structural and floristic components of the plant communities, by indicator plant species, by the age–growth curves of the tree species and by factors which are related to the macro- and microclimate, the soil types, and landforms. Thus phytogeomorphology is an important planning and management tool in forestry; but quite surprisingly many professional foresters who are trained and experienced in using the age–growth curves of the tree species, soil types, and the natural vegetation component of phytogeomorphology often have little or no experience of the geomorphic component for assisting in site quality evaluation and in the stratification of the natural forest into more homogeneous types ahead of field inventory. Phytogeomorphic stratification is useful for the subdivision of the dominant forest species into smaller groups prior to aerial photographic interpretation or for assistance in the choice of forest tree species for the afforestation of abandoned agricultural lands, or in the choice of natural forest to be cleared for forest plantations or for agriculture.

In the humid tropics it is often extremely difficult to stratify rainforest into homogeneous units based on differences in the vegetation, because there are many species (possibly 100 different rainforest tree species or more per hectare) and a profuse uniformity of the forest physiognomy. On the other hand, stratification based on landforms can be meaningful and may provide recognizable land systems and also several distinct forest land facets within each land system.

Even in a cool temperate climate with relatively few tree species, the land unit approach can be useful. Thus in Brechfa Forest (Wales) there is a noticeable difference in site quality for Sitka spruce (*Picea sitchenis*) and Norway spruce (*Picea abies*) on two adjoining land regions (Silurian and Ordovician sediments). Again, under cool temperate conditions, difficulty in assessing the site quality from the natural vegetation may be increased by finding different stages of vegetation succession on certain sites and a study of the landforms may be more meaningful to long-term planning. In Sweden, for example, Scots pine and Norway spruce may each occur separately on similar land facets because of being at different stages in the succession of the forest toward its climatic climax. Under different circumstances, however, the vegetation may be the more important in identifying the site quality. In cool temperate western Europe, the landform of two alluvial land facets may suggest similar site qualities, but the dominance of *Molinea coerulea* on one will suggest free-flowing subsurface water and the suitability for afforestation with Sitka spruce or Norway spruce, whereas on the other land facet, *Sphagnum* and the grass species *Nardus stricta* will indicate a difficult site requiring drainage of the stagnant water before afforestation with *Pinus contorta*.

In the warm temperate sclerophyllous forests of Mount Disappointment in southeastern Australia, research has indicated the potential value of a phytogeomorphic approach both for forest inventory and forest management. The forest is dominated by three land systems covering areas of 50 to 100 km^2. Although there is a distinct south–north rainfall gradient of 1400 to 800 mm across the plateau, the transition from wet sclerophyllous forest on granodiorite in the south to dry sclerophyllous forest on a deep regolith over Devonian sediments in the north is too sudden to be primarily due to rainfall and is correlated with the change in land systems.

The boundaries of the three land systems identifiable on aerial photographs provided useful strata for forest inventory. The granodiorite land system contains the most valuable timber species *Eucalyptus regnans* (mountain ash) and the tree species on the exposed folded Devonian sediments contain relatively few trees reaching timber size. Also, for purposes of aerial photographic interpretation, the strata based on land systems enabled the 13 tree species within the forest to be divided into groups of three, five, and seven. Furthermore, as the physiognomy of the vegetation of each of the three landforms is distinct, each has its own value for native fauna management and recreational purposes.

13.11   RANGELAND MANAGEMENT

Rangeland is generally located in areas where, because of low rainfall, low temperatures, or both, natural tall woody vegetation is relatively sparse, arable agriculture is difficult or impossible, and pastoralism the most suitable land use. Land values are usually too low for much improvement to be economic and reliance must be placed on low stocking rates and management of the vegetation to conserve and permit regeneration of the more nutritious and palatable species and prevent erosion. Rangeland in the Old World is generally used for nomadic herding and in the New World, Australasia, and South Africa, is used for extensive commercial ranching. Phytogeomorphic surveys are often the recommended first step in the appraisal of such areas.

Under the climatically marginal conditions characteristic of rangelands, the vegetation closely reflects the minor variations in site conditions and quickly reacts to changes in temperature or rainfall. In semiarid lands, valleys, basin sites, and permeable soils stand out as areas favorable for increased production of biomass. In subarctic areas, local improvement of drainage by deepening of the permafrost layer and slopes with favorable solar exposure are likewise associated with enhanced productivity. A phytogeomorphic interpretation not only permits a separation of such areas, but also can with ground survey allow their relative values to be quantified.

An example of the use of landscape units for classifying range quality is the survey of Jordan carried out by the Hunting Technical Services in 1956. A plant ecologist and a geologist, using aerial photographic interpretation and field traverses, divided the country into areas of different range types that, because of their uniformity of climate, soil, and vegetation, could be submitted to uniform land use. The major classes were grassland, brush, bare rock, deciduous or coniferous forest, and bake ground (mud flats), each of which was subdivided into classes that generally reflected the major regions of the country.

The results formed a useful basis for planning. They indicated the areas of the Mediterranean and steppe zones where overgrazing or agricultural exploitation poses the greatest danger to erosion. They also indicated where irrigation would be possible, and where the supply of flushes of grass by water spreading or the provision of water points in areas with grazing potential would be beneficial.

13.12   MIGRATORY PESTS

Both the landforms and the vegetation contribute to the development and movements of certain pests, of which the locust is probably the best known. Locusts threaten crops and rangeland over some 30 million km^2 in 60 countries with more than a fifth of the world's population. Build-ups of locust populations are dependent on meteorological and phytogeomorphic conditions and their monitoring and control require both factors to be considered. Vegetation development in the arid natural habitat of the desert locust (*Schistoceria gregaria*) can trigger a rapid population increase, which, if uncontrolled, becomes a plague. At its height the plague is characterized by large, highly mobile swarms, devasting crops in semiarid to temperate climatic zones.

The prediction and control of locust swarms are supplemented by analyzing meteorological data and by mapping the development of biomass density over these wide areas on a regular basis, which includes the use of satellite imagery (FAO, 1982). In addition, the landform and the density of the vegetation in these arid areas closely reflect the increasing downstream concentration of soil water in identifiable land facets, beginning at the tips of the tributary channels and attaining maximum concentration in the floors of the main trunk streams.

13.13 TERRAIN CLASSIFICATION FOR ENGINEERING

Both landforms and the vegetal cover are important to the classification of the landscape for engineering purposes. In addition, information is needed on the soil association, its textural type, subsurface water conditions, and the type and density of the drainage. In studying the landforms attention is paid to the lithology of the area, its surface expression, and the microtopography, including rock outcrops and slope characteristics.

An example of a land unit approach to terrain classification for engineering is provided by the 1970 program of CSIRO in Victoria, Australia (Grant, 1972). The basis of the program was that any area of land contains valid naturalistic terrain classes that can be uniquely defined in terms of topography (slope characteristics, underlying lithologic, and tectonic characteristics), soils, and vegetation. The classification covering an area of about 14,000 km^2 was made at four levels, which correspond closely to the land unit classification described earlier.

These levels are provinces (land regions), terrain patterns (land systems), terrain units (land facets), and terrain components (land elements). The importance of constant lithology is emphasized by grouping widely dispersed terrain patterns on similar rock type into the same provinces. The final map using a numerical system of nomenclature provides 16 provinces on Devonian, Silurian, and Tertiary silts, shales, sandstones, granodiorite, acid volcanics, Quaternary sands, and younger basalts. Based on repetition of terrain units and variations in topography, including amplitude, dissection status and type, and density of drainage patterns, these are divided into 60 terrain patterns. The terrain units within each terrain pattern were identified by drainage lines, slope, soil associations (sands, clay, silts, and organic), and plant formations–plant subformations.

13.14 TOWN AND COUNTRY PLANNING

Phytogeomorphology enters into both urban and rural planning, but the scale and emphasis are different between them. In urban areas land values are higher, site developments more intensive, and the visual scene is mainly architectural. The business districts of cities and industrial complexes, from which settlement radiates, generally require flat, level, dry sites with bedrock deep enough below the surface to avoid constructional and drainage problems. Residential property prefers somewhat steeper gradients and is better able to accommodate to smaller and less compact sites. Land that has become urbanized around sites of mineral extraction is often exposed to unsightly environmental damage. It is clearly desirable in such areas to use photogeomorphic data to evaluate the site for development, to maximize the conservation and enhancement of the natural sites, and to minimize and ameliorate the ill effects of urbanization and the scars left by extractive processes.

In rural areas views are wider and recreational uses less intensive, but contrasts in land values can be as extreme. Areas that historically have had little economic value may suddenly experience spectacular increases as a result of tourism, especially when this involves the development of leisure and retirement centers in locations selected for their climate and landscape, including proximity to mountains, lakes, and seacoasts. This effect can be seen, for instance, in the Sunbelt of the United States and in resort areas of the Rockies and the Swiss Alps.

Rural land is used not only for plant and animal production but also fulfills a variety of recreational needs that differ in the nature and intensity of their landscape requirements. Urban parks experience especially intensive use. Playing fields need to be firm, level, and well drained, and the sward resistant to trampling. Public gardens and cemeteries have somewhat less exacting surface requirements, but both should be selected with due consideration of the phytogeomorphic factor and should be amenable to the planning of specialized vegetation and the public gardens to the creation of ponds. Land for scientific and educational purposes such as nature reserves is generally either preserved because it is typical of an area or because it contains some exceptional habitat of plant or animal species. The main concern is the protection of the whole phytogeomorphic environment in order to ensure the survival of the required species. Finally, areas for wider-ranging forms of sport and recreation are selected mainly on the basis of their esthetic properties with emphasis on the presence of hills, forests, seacoast, lakes, and running water. Such land provides nature trails for rambling and picnicking, bridle paths for running and riding, and open water for swimming, boating, and fishing. Other sports have specialized needs, that is, undulating areas for golf and parklands, and for aerial sports such as hang-gliding, parascending, gliding, and ballooning open areas that provide appropriate aerodynamic conditions.

Special attention must also be paid to land required for transport purposes such as roads, railways, canals, and airports because of their long narrow sites and the problems posed by the need to harmonize the interest of abutting owners and travelers. Although highways provide the means of bringing people into contact with the countryside, they also bring danger, noise, and pollution to their surroundings. Road alignments should therefore avoid traversing land with either important physical obstructions or with high social value in terms of historical, cultural, or scenic considerations.

Practical projects usually derive their phytogeomorphic data from a synthesis of existing thematic maps. An example involving a wide variety of types of land development is given by Froehlich et al. (1978).

Lehigh, Virginia was a new urban community planned and developed in an undeveloped 1000-acre rural site in a gently undulating grassy upland area capped by gravel and scarred by abandoned gravel pits 10 miles southwest of the District of Columbia. Its margins are wooded ravines and it commands good views to the Appalachian foothills and the Potomac estuary.

Topographic, geologic, and hydrologic maps provided the basic data and served as sources for the compilation of other *derivative* maps. The landform map showing slope categories was derived from the topographic map and surface material and gravel subcrop maps from the geological map.

By combining these derivative maps, *interpretative* maps, such as those of slope stability, were generated. Synthesizing derivative maps with selected basic data gave *capability* maps, which provided possible solutions to land-use planning problems. At the same time, planners considered other aspects such as air quality, noise, wildlife habitats, and vegetation types to determine land-use suitability for the area. This environmental input was in turn incorporated with such other factors as adjacent land use, economics, housing needs, transportation, and public facilities to reach final land use recommendations.

FOURTEEN

CONCLUSIONS

This book draws attention to the importance of four major environmental factors (climate, landform, vegetation, and soil) as influencing the development of the natural landscape. Also, it has been pointed out that each of these factors separately expresses identifiable characteristics of the environment and contributes to the survey and mapping of the landscape. Sometimes vegetal differences or landform differences or their combination are the most important, whereas in other circumstances soil differences or climatic differences may predominate.

As explained previously, the earth's surface at the broadest level of mapping is most readily divided into zones based primarily on climatic differences and hence the importance of macroclimate in identifying agroecological zones. However, as we move to more detailed surveys and mapping at scales larger than about 1:5 million, the importance of recognizing and delineating landforms, including their rock type, slope, topographic expression and drainage patterns, and the selection of parameters representing the characteristics of landform units, increases rapidly. In the text emphasis is placed on landforms and vegetation in studying the landscape; and this is substantiated by the premise that aided by modern technology, particularly remote sensing analysis, these two factors are the most readily observed continuously over large areas of the earth's surface.

As we progress to study the landscape in detail and mapping at larger scale (1:25 000), the importance of recognizing subtle changes in the vegetation increases and the skills of the plant ecologist are needed, first to discriminate

198

physiognomic differences and possibly later, floristic communities. The interpretation of subtle changes in the characteristics of vegetation is particularly useful in assessing the combined effect of two or more environmental factors, the significance of which may be difficult to integrate using computer-assisted algorithms.

In the text plant–landform relations have been studied within the context of the scalar hierarchy of phytogeomorphic units considered in Chapters 5 and 6. These phytogeomorphic units are defined in terms of a number of attributes rather than by a single one, that is, they are polythetic rather than monothetic and boundaries on maps therefore need to be a synthesis of these attributes. They range from the broad linkages between plant panformations and land zones at one extreme through intermediate units (e.g., land system) to the detailed interconnections of small plant communities with the soil inside single land elements at the other. Research has shown that the land facet is especially valuable because of its internal homogeneity, the land system because of its usefulness in mapping at intermediate scales and as the largest land unit useful to land management, and the land province because of the apparent accordance of the physiography with practically significant vegetation formations and climatic zones. At all scales phytogeomorphic units tend to be recurrent and provide a useful vehicle for recognizing similarities between analogous areas. This can lead to the transfer of scientific and practical data between them and to the identification and explanation of dissimilarities and anomalies. It is sometimes advantageous to group contiguous land units that are relevant to management practices and to designate them management units; but this is not feasible where differences between neighboring land units are reflected in contrasting land uses.

The main value of applying phytogeomorphology to studies of the earth's surface is its contribution to studies of land potential and land use. This derives from two main factors: the practical homogeneity provided by phytogeomorphology, especially at the level of meso- and microunits, and their ready recognition on remotely sensed imagery. This value of remote sensing is recognized in most environmental disciplines, and the wide usefulness of such imagery plays an important role in integrating the studies provided by the different disciplines. Such advances have in their turn strongly influenced phytogeomorphic studies by facilitating thematic interpretations, and by refining the direction of ground sampling programs upon which the collection of basic environmental data depends. The contribution of remote sensing to studies of the landscape also favors the teaching of phytogeomorphology as a single subject.

The specific contribution of phytogeomorphology to survey is viewed not to replace but to assist other forms of natural resource activities because of its synthesizing value in multidisciplinary studies, its retention of a constant theme over a range of scales, and the relative ease with which its units may be observed in the field. Thus first, phytogeomorphology provides, at relatively low cost, thematic base maps of the land surface for both general and applied purposes of value to a range of different users. Second, the hierarchical scheme of mutually related land units provides a framework that can, with a constant theme of land-form–vegetation relations, answer questions at a wide variety of scales. A phytogeomorphic study can, for example, at the same time show (a) the broad regional context of a project area, (b) its general character, and (c) provide highly detailed data on selected small sites. Generally it can be stated that the combined emphasis of landforms and vegetation has relevance in two ways. First, these contribute to the integrated appraisal of land use and static natural resources both in the past, for example, archeological time, and for present-day planning. Second, they provide information on land potential and lead to the prediction and the monitoring of destructive events. These may be either direct natural hazards such as floods, fires, and earthquakes or environmentally dependent pests.

For many purposes it is now preferable to digitize location-specific data so that it can be handled by computer. This increases the speed and flexibility in processing and presenting geographically referenced information. It permits the data to be manipulated in a number of ways before being presented in numerical, diagrammatic, or cartographic form. It is possible, for instance, automatically to make summaries of data, scale changes, selective searches, and aerial measurements, and to perform such operations as automatic contouring or to combine and compare two or more different layers of data. This is important in phytogeomorphic studies because it permits, for instance, a layer of data newly acquired from remote sensing to be combined or compared with earlier layers from geological, geomorphic, or vegetation maps.

When viewed at regional, national, or international scale, the large amount and variety of environmental data and the wide range of locations to which it relates pose problems of data management which may best be solved by geographic information systems. Where these are relatively simple, they can be operated manually, but for large or complex areas, computer-assisted methods become necessary. This is leading to the development of geographic information centers that can process the data for the benefit of users. A number of these centers are being developed for resource data by national and provincial bodies in different parts of the world, but with a few exceptions the systems they use are still experimental rather than fully operational.

In the future the continued refinement of modern technology can be expected to create more and more data through remote sensing and to provide more usable information through computer processing associated with data banks and geographic information systems. Already remote sensing is generating huge quantities of phytogeomorphic data stored as images on aerial film or on digital tapes from which satellite or radar imagery can be created. In fact, much of this phytogeomorphic data are already only partly used and must await major advances in computer processing techniques or the extensive training and reemployment of photographic interpreters. In years gone by, when remote sensing data were not so readily available, observations on landforms were much more difficult and often neglected in favor of field-collected data on soils and vegetation.

With the recent interest in geographic information systems, which is being stimulated by the rapid development of inexpensive minicomputers, there is an added awareness of the need for comprehensive information on the earth's surface. This can be expected to increase the demand for phytogeomorphic data, much of which, as already mentioned, are recorded on airborne and satellite imagery, but it will also draw attention to areas in which information is lacking and urgently needed. These include information on changes in land use, land quality, desertification, and vegetal cover, which result from population pressures, soil degradation, possible climatic changes, the dearth worldwide of geomorphic maps, and information on coastal and inland waters.

As we know, in this rapidly changing world, new major water surfaces, observable on satellite imagery, are being created by the damming of major rivers (e.g., Nile, Yangtze, Zambesi, Volta) and the land use of water–vegetation climaxes are being irreversibly changed (e.g., mangrove forest to fish culture to rice production). Urban areas are being newly created on highly productive agricultural lands. Where only a few years ago shifting cultivation (taungya) was a traditional and sound agricultural practice, the period of bush fallow is being reduced (e.g., in west Africa, formerly 15–20 yr, now 3–5 yr). In regions where rainfall is marginal (e.g., Sahel) animal and human carrying capacities are being far exceeded, with the result that desertlike conditions are being created. In all these circumstances there is an important role in survey, management, and planning for phytogeomorphology.

REFERENCES

AGARD (North Atlantic Treaty Organization Advisory Group for Aerospace Research and Development) (1970) *Information Analysis Centres*, AGARD Conference Proceedings, No. 78, Schiphol, Amsterdam.

American Society of Photogrammetry (1960). *Manual of Aerial Photographic Interpretation*, Falls Church, Virginia.

American Society of Photogrammetry (1975, 1982). *Manual of Remote Sensing*. (2 vols.), Falls Church, Virginia.

Anderson, J. R., E. E. Hardy, J. T. Roach, and R. E. Witmer (1976). *A land use and land cover classification system for use with remote sensor data*, USGS Professional Paper 964, Washington, D.C.

Anstey, F. (1960). *Digitized Environmental Data Processing*, Headquarters Quartermaster Research and Engineering Command, Research Study Report RER-31, U.S. Army, Natick, Massachusetts.

Avery, T. E. (1964). To stratify or not to stratify, *Journal of Forestry* 62, 106–108.

Baboulene, B. (1969). *Critical Path Made Easy*, Duckworth, London.

Ballew, G. I. (1975). Correlation of Landsat-1 multispectral data with surface geochemistry, *Proceedings of 10th Annual Symposium on Remote Sensing of the Environment* (Vol. 2), Ann Arbor, Michigan, pp. 1045–1055.

Barrett, E. C. (1974). *Climatology from Satellites*, Methuen, London.

Beadle, N. C. and A. B. Costin. (1952). Ecological classification and nomenclature, *Proceedings of the Linnean Society of NSW* 77, 61–82.

Beard, J. S. (1944). Climax vegetation in tropical America, *Ecology* 25, 127–158.

Beckett, P. H. T. and R. Webster (1965) *A Classification System for Terrain*, Military Engineering Experimental Establishment, Report No. 872, Christchurch, Dorset.

Becking, R. W. (1959). Forestry applications of aerial color photography, *Photogrammetric Engineering* 25, 559–565.

Belcher et al. (1951) *An Airphoto Analysis Key for the Determination of Ground Conditions*, Vol. 1, U.S. Office of Naval Research, Cornell University, Ithaca, New York.

Bendelow, V. C. and R. Hartnup (1980). *Climatic Classification of England and Wales*, Soil Survey Technical Monograph No. 15, Harpenden.

Bie, S. W. (Ed.) (1975). Soil information systems, *Proceedings of the Meeting of the ISSS Working Group on Soil Information Systems*, Wageningen, The Netherlands.

Black, C. A. (Editor-in-chief) (1965). Methods of soil analysis, *Agronomy* **9,** American Society of Agronomy, Madison, Wisconsin.

Blasco, F. (1984) Personal communication.

Bourne, R. (1931). *Regional Survey and Its Relation to Stocktaking of the Agricultural and Forest Resources of the British Empire*. Oxford Forestry Memoir No. 13, Clarendon Press, Oxford.

Bowman, I. (1914). Forest physiography, physiography of the U.S. and principal soils in relation to forestry, Wiley, New York

Braun-Blanquet, J. (1932). *Plant Sociology*, McGraw-Hill, New York.

Braun-Blanquet, J. (1947). *Carte des Groupements Végétaux de la France, Région Nord-Ouest de Montpellier*, Station International de Geobotanique Mediterraneenne et Alpine, France.

Braun-Blanquet, J. (1964). *Pflanzensoziologie: Grudzüge der Vegetationskunde*, 3rd. edition, Springer-Verlag, Wien-New York.

Bray, J. R. and J. T. Curtis (1957). An ordination of the upland forest communities of southern Wisconsin, *Ecol. Monographs* **27,** 325–349.

Brink, A. B., J. A. Mabbutt, R. Webster, and P. H. T. Beckett (1966). *Report of the Working Group on Land Classification and Data Storage*, Military Engineering Experimental Establishment, Christchurch, Dorset.

Brooks, R. R. (1972). *Geobotany and Biogeochemistry in Mineral Exploration*, Harper & Row, New York.

Brooks, R. R. (1983). *Biological Methods of Prospecting for Minerals*, Wiley, New York.

Bucknell, J. (1952). Some methods of classifying climates, M.Sc. thesis, University of Reading.

Büdel, J. (1963). Klima-genetische geomorphologie, *Geographische Rundschau* **15,** 269–285.

Budyko, M. I. (1956). *The Heat Balance of the Earth's Surface* (trans. by N. I. Stepanova), U.S. Weather Bureau, Washington, D.C., 1958.

Cadigan, R. A., L. R. Ormsbee, R. A. Palmer, and P. T. Voegeli (1972). *Terrain Classification: A Multivariate Approach*, U.S. Geological Survey Report USGS-GD-72-024, Washington, D.C.

Cannon, H. L. (1960). *The Development of Botanical Methods of Prospecting for Uranium on the Colorado Plateau*, Geological Survey Bulletin 1085-A, Washington, D.C.

Christaller, W. (1966). *Central Places in Southern Germany*, translation of 1933 (Jena) edition by C. W. Baskin, Pentrice-Hall, Englewood Cliffs, New Jersey.

Christian, C. S. and G. A. Stewart (1968). Methodology of integrated surveys, *Proceedings of UNESCO Conference on Aerial Surveys and Integrated Studies*, Toulouse (UNESCO, Paris).

Clarke, G. (1964). *Archeology and Society* (3rd ed.), Methuen, London.

Clarke, G. R. (1957). *The Study of the Soil in the Field* (4th ed.), Clarendon Press, Oxford.

Clements, F. E. (1916). *Plant Succession*, Publication 242, Carnegie Institution of Washington.

Coaldrake, J. E. (1961). *The Ecosystems of the Coastal Lowlands (WALLUM) of Southern Queensland*, CSIRO Bulletin 283, Melbourne.

Cochran, W. G. (1977). *Sampling Techniques* (3rd ed.), Wiley, New York.

Cole, J. P. and C. A. M. King (1968). *Quantitative Geography*, Wiley, London.

Conklin, H. E. (1959). The Cornell system of economic land classification, *Journal of Farm Economics* **41,** (1), 548–557.

Cooke, R. U. and A. Warren (1973). *Geomorphology in Deserts*, Batsford, London.

Cousins, L. B. (1970). The Harwell heat transfer and fluid flow information analysis center. In *AGARD Information Analysis Centres*, North Atlantic Treaty Organization Conference Proceedings, No. 78, London **4**, 1-6.

Cowan, D. J., J. N. Bayne, and D. A. Fairey (1976). *Development and Applications of the South Carolina Computerized Land Use Information System*, South Carolina Land Resources Commission.

Croizat, L. (1952). *Manual of Phytogeography*, Uitgeverij Dr. W. Junk, The Hague.

Cuanalo, de la, C. H. E. and R. Webster (1970). A comparative study of numerical classification and ordination of soil profiles in a locality near Oxford. I. Analysis of 85 sites, *Journal of Soil Science* **21**, 340-352.

Curtis, J. T. (1959). *The Vegetation of Wisconsin: An Ordination of Plant Communities*, University of Wisconsin Press, Madison.

Dale, E. and L. C. Michelon (1966). *Modern Management Methods*, Penguin Books, Harmondsworth, United Kingdom.

Dansereau, P. (1951). Description and recording of vegetation upon a structural basis, *Ecology* **32**, 172-229.

Demek, J., C. Embleton, J. F. Gellert, and H. Th. Verstappen (eds.) (1972). *Manual of Detailed Geomorphological Mapping*, Academia, Prague.

Dent, F. (1980). Private communication.

Dickinson, R. E. (1930). The regional functions and zones of influence of Leeds and Bradford, *Geography* **15**, 548-557.

Driggers, W. G., J. M. Downs, J. R. Hickman, and R. L. Packard (1978). Data base design for a worldwide multicrop information system. In *NASA, the LACIE Symposium: Proceedings of Technical Sessions*, NASA Johnson Space Center, Houston, Texas, pp. 1085-1096.

Driscoll, R. S., D. H. Betters, and H. D. Parker (1978). Land classification through remote sensing—techniques and tools, *Journal of Forestry* 656-661.

Dudal, R. (1982). Land resources and production potential for a growing world population: Optimizing yields—the role of fertilizers, *Proceedings of the 12th Congress of the International Potash Institute*, Berne, pp. 277-288.

Dudal, R., G. M. Higgins, and A. H. Kassam (1982). Land resources for the world food production, *12th International Congress of Soil Science*, New Delhi.

Du Rietz, G. E. (1936). Classification and nomenclature of vegetation units, *Svensk Botanisk Tidskrift* **30**, 580.

ESRI (Environmental Systems Research Institute) (1979). *Bernardino Forest Wildland Recreation Study*, Redlands, California.

Experimental Cartography Unit, Royal College of Art (1970). Recent advances in automatic cartography. In *Papers for discussion at Symposium on New Possibilities and Techniques for Land Use Surveys with Special Reference to the Developing Countries*, convened jointly by Shell International Petroleum Company and World Land Use Survey Commission of International Geographical Union, Chairman Professor H. Boesch, Shell Centre, London, K1-K6.

FAO (1978). Report on the Agro-Ecological Zones Project, World Soil Resources Report 48, Vol. 1. *Methodology and Results for Africa*, Vol. 2, *Results for Southwest Asia*, Project Coordinator, G. M. Higgins, Rome.

FAO (1979). *Report on Agriculture Towards 2000*, Rome.

FAO (1981). *Towards the Year 2000*, Rome.

FAO (1982). *Remote Sensing Applied to Renewable Resources*, Remote Sensing Centre, Rome.

FAO-UNEP-UNESCO (1981). *A Provisional Methodology for Soil Degradation Assessment*, FAO, Rome.

FAO-UNESCO (1974). *Soil Map of the World at 1:5,000,000*, Vol. 1, Legend, Paris.

FAO-UNESCO (1981). *Soil Degradation Map of the World at 1:5,000,000*, Legend, Paris.

Fenneman, N. M. (1916, 1928). Physiographic divisions of the United States, *Annals Assoc. Amer. Geographers* **6**, 19-98. and **18**, 261-353.

Ferrel, W. (1856). Essai sur les vents et les courants de l'ocean, quoted in M. Brillouin (1900), *Memoires Originaux sur la Circulation Générale de l'Atmosphere, Paris*, and summarized in N. Shaw (1926), *Manual of Meteorology*, Vol. 1, *Meteorology in History*, Cambridge University Press.

Food and Agriculture Organization of the United Nations, see FAO.

Froelich, A. J., A. D. Garnaas, and J. N. Van Driel (1978). Franconia area, Fairfax County, Virginia. In G. D. Robinson and A. M. Spieker (Eds.), *Nature to Be Commanded*, U.S. Geological Survey Professional Paper 950, U.S. Government Printing Office, Washington, D.C.

Gardner, R. A. and A. E. Wieslander (1957). The soil vegetation survey in California, *Soil Science Society of America Proceedings* **21**, 103-105.

Gaussen, H. (1948). *Carte de la Végétation de la France*, Service de la Carte de la Végétation de la France, Toulouse.

Gaussen, H. (1955a). Les climats analogues a' l'échelle du monde, *Compte Rendu Acad. Agr. France* **41**.

Gaussen, H. (1955b). Expression des milieus par des formules écologiques, leur répresentation cartographique, *Annales Biologiques* **31**, 257-269.

German Research Society (1977). *Geography of Climates*, Afrika Kartenwerk, Tunisian and Ugandan areas at 1:1,000,000, Stuttgart.

Gibbons, F. R. and R. G. Downes (1964). *A Study of Land in Southwestern Victoria*, Soil Conservation Authority of Victoria, Australia.

Glentworth, R. L. and H. G. Dion (1949). The association of hydrologic sequence in certain soils of the podzolic zone of northeast Scotland. *J. Soil Sc.* **1**, 35-49.

Goodall, D. W. (1952) Quantitative aspects of plant distribution, *Biol. Rev.* **27**, 194-245.

Gorshkov, G. and A. Yakushova (1967). *Physical Geography* (trans. from the Russian by A. Gurevitch), Mir Publishers, Moscow.

Gradman, R. (1909). Über Begriffsbildung in der Lehre von Pflanzenformationen, *Bot. Jb. Beibl* **99**, 91-103.

Grant, K. (Ed.) (1968). *Proceedings of Study Tour and Symposium on Terrain Evaluation for Engineering*, CSIRO Division of Applied Geomechanics, Australia.

Grant, K. (1972). *Terrain Classification for Engineering Purposes of the Melbourne Area, Victoria*, CSIRO Division of Applied Geomechanics, Technical Paper No. 11, Australia.

Gray, T. I. and D. G. McCrary (1981). An application of Advanced Very High Resolution Radiometer data to monitor the world's agriculture, *Proceedings of the International Conference on Rice*, Hyderabad, India.

Great Britain, Ministry of Agriculture, Fisheries and Food (1976). *The Agricultural Climate of England and Wales*, Technical Bulletin 35, HMSO, London.

Gregory, S. (1978). *Statistical Methods and the Geographer* (4th ed.), Longmans, London.

Greig-Smith, P. (1964). *Quantitative Plant Ecology*, Butterworth, London.

Grigg, D. (1967). Regions, models and classes. In R. J. Chorley and P. Haggett (eds.), *Models in Geography*, Methuen, London, pp. 461-509.

Grisebach, A. H. R. (1872). *Die Vegetation der Erde nach ihrer Klimatischen Anordnung*, Leipzig.

Hadley, G. (1735). Concerning the cause of the general trade wind, *Phil. Trans. Roy. Soc. (London)* **39**, 58-62.

Haeckel, E. (1866). *Generale Morphologie der Organismen (Bd. 1)*, Berlin.

Hagberg, E. (1956). *Aerial Photographs for Mapping Purposes for Forests in Sweden*, I. S. P., Stockholm.

Hagood, M. J., N. D. Danilevsky and C. O. Beum (1941). An examination of the use of factor analysis in the problem of sub-regional delineation, *Rural Sociology* **6**, 216-233.

Harper, R. M. (1943). *Forests of Alabama*, Monograph 10, Geological Survey of Alabama.

Heath, G. R. (1956). A comparison of two basic theories of land classification and their adaptability to regional photo-interpretation key techniques, *Photogrammetric Engineering* **22**, 144-168.

Heinsdijk, D. (1960). Surveys applicable to exclusive forest areas (South America). *FAO Fifth World Forestry Congress*, Rome.

Henderson-Sellers, A. and N. Stockdale (1979). Changes in surface characteristics: A pilot survey for the British Isles using NOAA imagery. In J. A. Allan and R. Harris (Eds.), *Remote Sensing and National Mapping*, Remote Sensing Society, Reading, pp. 90-100.

Hettner, A. (1930). *Die Klimate der Erde*, Teubner, Leipzig.

Hicks, J. P. (1977). *Managing Natural Resources Data: Minnesota Land Management Information System*, Council of State Governments, Lexington, Kentucky.

Higgins, G. M. and A. H. Kassam (1981). FAO's agro-ecological approach to determination of land potential, *Pedologie* **XXXI**, 147-168.

Hills, G. A. (1961). *The Ecological Basis for Land-Use Planning*, Research Report No. 46, Ontario Department for Lands and Forests, Toronto.

Holdridge, L. R. (1947). Determination of world plant formations from simple climatic data, *Science* 105(2727), 367-368.

Holdridge, L. R. (1966). The life zone system, *Adsonia* **6**, 199-203.

Holdridge, L. R. and J. A. Toshi (1972). The world life zone classification system and forestry research, *Seventh World Forestry Congress*, 7CFM/C:V2G, FAO, Rome.

Holmes, A. (1956). *Principles of Physical Geology*. Thomas Nelson, Edinburgh.

Howard, J. A. (1959). The classification of woodland in western Tanganyika by type-mapping from aerial photographs, *Empire Forestry Review* **38**, 348-364.

Howard J. A. (1970a). *Aerial Photo-Ecology*, Faber and Faber, London.

Howard, J. A. (1970b). Steroscopic profiling of land units from aerial photographs, *Australian Geographer* **11**(3), 259-68.

Howard, J. A. (1970c). Stereoscopic profiling and the photogrammetric description of woody vegetation, *Australian Geographer* **11**(3), 359-372.

Howard, J. A. (1970d). Multi-band concepts of forested land units. In *International Symposium of Photo-Interpretation* (Vol. 1). International Archives of Photogrammetry, Dresden.

Howard, J. A. (1976). Satellite remote sensing of agricultural resources for developing countries, *The Colston Symposium: Remote Sensing of the Terrestrial Environment*, University of Bristol, United Kingdom.

Howard, J. A. and C. W. Mitchell (1980). Phyto-geomorphic classification of the landscape, *Geoforum* **11**, 85-106.

Howard, J. A. and J. Schade (1982). *Towards a Standardized Hierarchical Classification of Vegetation for Remote Sensing*, Remote Sensing Centre, FAO, Rome.

Howard, J. A. and D. C. Schwaar (1978). *Role and Application of High Altitude Aerial Photography in the Humid Tropics with Special Reference to Sierra Leone*, FAO, Rome.

Howard, J. A. and A. van Dijk (1980). Towards a world index of space imagery, paper presented to *International Society of Photogrammetry*, Commission VII, Rome.

Huang, W. T. (1962). *Petrology*, McGraw-Hill, New York.

Hunting Technical Services, Ltd. (1956). *Report on the Range Classification Survey of the Hashemite Kingdom of Jordan*, for Government of Jordan, London.

Hunting Technical Services, Ltd. (1963). *Roseires Soil Survey*, Report No. 1 Gezira Extension Area Soil Survey and Land Classification, Republic of the Sudan, Ministry of Agriculture, Vol. 1, Sir M. Macdonald and Partners, London.

Husch, B. (1963). *Forest Mensuration and Statistics*, Ronald Press, New York.

ICSU (International Council of Scientific Unions) (1979). *Fourth Consolidated Guide to International Data Exchange Through the World Data Centers*, Secretariat of the ICSU Panel on World Data Centers, Washington, D.C.

Jenkin, R. N. and M. A. Foale (1968). *An Investigation of the Cocoanut Growing Potential of Christmas Island* (Vols. 1 and 2), Directorate of Overseas Surveys, Land Resources Study 4, Tolworth.

Jenny, H. (1941). *Factors of Soil Formation*, McGraw-Hill, New York.

Joerg, W. L. G. (1914). Natural regions of northern America, *Annals Assoc. Amer. Geographers* **4**, 55–83.

Joly, F. (1962). *Études sur le Relief du Sud-est Marocain*, Travaux de l'Institut Scientifique Chérifien, Sér Géol. et Géog. Phys. Rabat.

Kemp, R. C., C. G. Lewis, and C. R. Robinson (1925). *Air Photo Survey and Mapping of the Forests of the Irrawaddy Delta*, Burma Forestry Bulletin No. 11.

King, L. J. (1969). *Statistical Analysis in Geography*, Prentice-Hall, Englewood Cliffs, New Jersey.

Kloosterman, B. and J. Dumansky (1978). Data management capabilities of the Canada Soil Information System. In A. N. Sadovsky and S. W. Bie (Eds.), *Development in Soil Information Systems*, Proceedings of the Second Meeting of the ISSS Working Group on Soil Information Systems, Varna/Sofia, Bulgaria, Wageningen, The Netherlands.

Knapp, E. M. and R. Ryder (1979). Automated geographic information systems and Landsat data: A survey. In *Computer Mapping in Natural Resources and the Environment*, pp. 57–68. Volume 4 of Harvard Library of Computer Graphics in 1979 Mapping Collection, Laboratory for Computer Graphics and Spatial Analysis, Harvard University.

Köppen, W. (1900). Versuch einer Klassifikation der Klimate, Vorzugsweise nach ihren Beziehungen zur Pflanzenwelt, *Geographische Zeitschrift* **6**.

Köppen, W. (1923). *Die Klimate der Erde: Grundriss der Klimakunde*, Walter de Gruyter, Berlin; summarized by Committee of the Geographical Association in *Geography*, **22**(4), 253–282 (1937).

Köppen, W. (1931). *Grundriss der Klimakunde*, De Gruyter Verlag, Berlin.

Krumbein, W. C. and F. A. Graybill (1965). *An Introduction to Statistical Methods in Geology*, McGraw-Hill, New York.

Küchler, A. W. (1967). *Vegetation Mapping*, Ronald Press, New York.

Lamb, H. H. (1972). *Climate: Past, Present and Future* (Vol. 1), Methuen, London E.C.4.

Lambert, J. M. and W. T. Williams (1962). Multivariate methods in plant ecology. IV. Nodal analysis, *J. Ecol.* **50**, 775–802.

Langston, R. P. (1970). Maritime pollution. In *AGARD, Information Analysis Centres*, North Atlantic Treaty Organization, Conference Proceedings No. 78, London.

Leopold, L. E., M. G. Wolman, J. P. Miller (1964). *Fluvial Processes in Geomorphology*, Freeman, San Francisco.

Lillesand, T. M. and R. W. Kiefer (1979). *Remote Sensing and Image Interpretation*, Wiley, New York.

Linton, D. L. (1951). The delimitation of morphological regions. In L. D. Stamp and S. W. Wooldridge (Eds.), *London Essays in Geography*, Longmans, London, pp. 199–218.

Lobeck, A. K. (1939). *Geomorphology: An Introduction to the Study of Landscapes*, McGraw-Hill, New York.

Loetsch, F. (1957). A forest inventory in Thailand, *Unisylva* **11**, 174–180.

Long, G. (1974). *Diagnostic Phyto-Écologique et Amenagement dux Territoire*, Collection d'Ecologie 4 (Vol. 1), *Principes Généraux et Methodes*, Masson et Cie. (Editeurs), Paris.

Losee, S. T. (1951). Photographic tone in forest interpretation, *Photogrammetric Engineering* **17**, 785–799.

Mabbutt, J. A. (1977). *Desert Landforms*, Vol. 2 of *An Introduction to Systematic Geomorphology*, ANU Press, Canberra.

McNeil, G. (1967). *Terrain Evaluation: Data Storage*, MEXE (Military Engineering Experimental Establishment), Technical Note 5/67, Christchurch, Hants.

Marchesini, E. and A. Pistolesi (1964). Landform maps of intermediate scale, *XX International Geographical Congress*, Section IX, London.

Maycock, P. F. and J. T. Curtis (1960). The phyto-sociology of boreal conifer–hardwood forests of the Great Lakes region, *Ecol. Monogr.* **30**, 1–35.

Mensching, H. (1950). Verbreitungskarten von Pflanzengesellschaften als Hilfsmittel für den Morphologen am Beispiel des Wesertales, *Mitteilungen der floristisch-soziologischen Arbeitsgemeinschaft*, N.F. Heft 2, Stolzenau an der Weser.

Michigan, University of (1979). *Comprehensive Resource Inventory and Evaluation System*, CRIES Special Reports 1–5, Ann Arbor.

Miller, R. L. and S. J. Kahn (1962). *Statistical Analysis in the Geological Sciences*, Wiley, New York.

Mills, H. L., et al. (1963) *Quantitative Physiognomic Analysis of the Vegetation of the Florida Everglades*, U.S. Corps of Engineers, Marshall University, West Virginia.

Milne, G. (1935). Some suggested units of classification and mapping particularly for East African soils, *Soil Research* **4**, 183–198.

Mitchell, C. W. (1973). *Terrain Evaluation*, Longmans, London.

Mitchell, C. W. and J. A. Howard (1978a). *Final Summary Report on the Application of Landstat Imagery to Soil Degradation Mapping at 1 : 5,000,000*, AGLT Bulletin 1/78, FAO, Rome.

Mitchell, C. W. and R. M. S. Perrin (1966). The subdivision of world hot deserts into physiographic units, *Actes du Ile Symposium International de Photo-Interpretation*, I.S.P. Commission VII, Sorbonne, Paris, IV-1, pp. 89–98.

Mitchell, C. W., R. Webster, P. H. T. Beckett, and B. Clifford (1979). An analysis of terrain classification for long range predictions of conditions in deserts, *Geographical Journal* **145**, 72–85.

Mitchell, C. W. and S. G. Willimott (1974). Dayas of the Moroccan Sahara and other arid areas, *Geographical Journal* CXL, 441–453.

Moessner, K. E. (1963). *A Test of Aerial Photo Classification for Forest Management*, U.S. Forest Service Research Paper 3.

Morrison, D. F. (1976). *Multivariate Statistical Methods*, ed. 2, McGraw Hill, New York.

Mueller-Dumbois, D. and H. Ellenberg (1974). *Aims and Methods of Vegetation Ecology*, Wiley, New York.

NOAA (National Oceanographic and Atmospheric Administration) (1983). *Global Vegetation Index Users Guide*, Satellite Data Services Division. Washington, D.C.

Northern Ireland Ordnance Survey (1918). *Mourne Mountains Outdoor Pursuits Map, 1:25,000*, Belfast.

Norwine, J. and D. H. Greegor (1983). Vegetation classification based on advanced very high resolution radiometer (AHVRR) satellite imagery, *Remote Sensing of Environment* **13**, 69-87.

Odum, H. T. (1971). *Environment, Power, and Society*, Wiley-Interscience, New York.

Ollier, C. D. (1969). *Weathering*, Oliver and Boyd, Edinburgh.

Paarma, H., H. Vartiainen, and J. Penninkilampi (1977). *Prospecting in Areas of Glaciated Terrain*, Institution of Mining and Metallurgy, London.

Palmen, E. (1951). The role of atmospheric disturbances in the general circulation, *Quarterly Journal Royal Meteorological Society* **77**, 337-354.

Passarge, S. (1919). *Die Grundlagen der Landschaftskunde*, L. Friederichsen, Hamburg.

Passarge, S. (1926). Geomorphologie der Klimazonen oder Geomorphologie der Landschaftsgürtel, *Petermanns Mitteilungen* **72**, 173-175.

Pearson, A. R. (1979). An integrated terrain analysis system. In *Proceedings of 13th International Symposium on Remote Sensing of the Environment* (Vol. 1), ERIM, Ann Arbor, Michigan, pp. 433-438.

Peltier, L. C. (1950). The geographic cycle in periglacial regions as it is related to climatic geomorphology, *Ann. Assoc. Am. Geog.* **40**, 214-236.

Peltier, L. C. (1962). Area sampling for terrain analysis, *Professional Geographer* **14**, 24-28.

Petrie, A. K., P. H. Jarrett, and R. T. Patten (1929). The vegetation of the Black Spur region—A study of the ecology of some Australian mountain eucalypt forests, *J. Ecol.* **17**, 223-248.

Piélou, E. C. (1977). *Mathematical Ecology*, Wiley-Interscience, New York: London

Poore, M. E. D. (1955). The use of phytosociological methods in ecological investigations. I. The Braun-Blanquet system, *J. Ecol.* **43**, 226-44.

Poore, M. E. D. (1963). *Advances in Ecological Research* (J. B. Cragg, Ed.), Academic, London.

Raisz, E. (1962). *Principles of Cartography*, McGraw-Hill, New York.

Raunkiaer, C. (1934). *The Life Forms of Plants and Statistical Plant Geography*, Oxford University Press, New York.

Richards, P. W. (1952). *The Tropical Rainforest*, Cambridge University Press, p. 399.

Robinson, A., R. Sale, and J. Morrison (1978). *Elements of Cartography*, Wiley, New York.

Rogers, E. J. (1961). Application of aerial photographs and regression techniques for surveying Caspian forests of Iran, *Photogrammetric Engineering* **27**, 811-816.

Rosayro, R. A. de, (1959). The application of aerial photography to stock mapping and inventories in rainforest in Ceylon, *Empire Forestry Review* **38**, 141-174.

Rossby, C. G. (1949). On the nature of the general circulation of the lower atmosphere. In G. P. Kuiper (Ed.), *The Atmospheres of the Earth and Planets*, University of Chicago Press, pp. 16-48.

Royal Meteorological Society, see Lamb (1972).

Rübel, E. (1930). *Pflanzengesellschaften der Erde*, Verlag Hans Huber, Bern-Berlin.

St. Onge, D. A. (1968). Geomorphic maps In R. W. Fairbridge, *Encyclopedia of Geomorphology*, Reinhold, New York.

Salus, R. M. (1981). *The State of the World Population, 1980*, UN Fund for Population Activities, New York.

Savigear, R. A. G. (1965). A technique of morphological mapping, *Annals Assoc. Amer. Geographers* **55**(3), 514–538.

Sayn-Wittgenstein, L. (1961). Recognition of tree species on air photographs by crown characteristics, *Photogrammetric Engineering* **27**, 798–819.

Schimper, A. F. W. (1903). *Plant Geography on a Physiological Basis*, Clarendon Press, Oxford; and *Pflanzengeographie auf Physiologischer Grundlage* (3rd ed.), Fischer Verlag, Berlin, 1898.

Schlesinger, J., B. Ripple, and T. R. Loveland (1979). Land capability studies of the South Dakota Automated Geographic Information System (AGIS). In *Computer Mapping in Natural Resources and the Environment*, Mapping Collection. Laboratory for Computer Graphics and Spatial Analysis, Harvard University.

Seale, R. S. (1975). *Soils of the Ely District* (sheet 173), Memoirs of the Soil Survey of Great Britain (England and Wales), Harpenden, Herts.

Severson, K. R. and J. F. Thilenius (1976). *Classification of Quaking Aspen Stands in the Black Hills and Bear Lodge Mountains*, U.S. Department of Agriculture Research Paper RM-166, Washington, D.C.

Smith, J. (1949). *Distribution of Tree Species in the Sudan in Relation to Rainfall and Soil Texture*, Sudan Government Ministry of Agriculture Bulletin No. 4, Khartoum.

Smyth, A. J. (1966). *The Selection of Soils for Cocoa*, Soils Bulletin 5, FAO, Rome.

Sneath, P. H. A. and R. R. Sokal (1973). *Numerical Taxonomy: The Principles and Practice of Numerical Classification*, Freeman, San Francisco.

Snedecor, G. and W. G. Cochran (1978). *Statistical Methods* (6th Ed.), Iowa State University Press, Ames.

Soil Survey of England and Wales (1977). *Winter Rain Acceptance Potential*, Ordnance Survey, Southampton.

Soil Survey Staff (1951). *Soil Survey Manual*, U.S. Department of Agriculture Handbook No. 18, Washington, D.C.

Sokal, R. R. and P. H. A. Sneath (1963). *Principals of Numerical Taxonomy*, Freeman, San Francisco.

Spurr, S. H. (1960). *Photogrammetry and Photo-Interpretation*, Ronald Press, New York.

Stamp, L. D. and E. C. Willatts (1934). *The Land Utilization Survey of Britain; an Outline of the First Twelve One-Inch Maps*, Land Utilization Survey of Britain, London School of Economics, London.

Stellingwerf, D. A. (1963). *Volume Determination on Aerial Photographs*, ITC Series B-18, Enschede, Netherlands.

Storie, R. E. (1964). Soil and land classification for irrigation development. *Proc. 8th Int. Cong. Soil Sci.* **V**, 873–882.

Strahler, A. N. (1964). Quantitative geomorphology of drainage basins and channel networks. Section 4-II of *Handbook of Applied Hydrology*, McGraw Hill Book Company, New York.

Strahler, A. N. (1969). *Physical Geography* (3rd ed.), Wiley, New York.

Sukachev, V. and N. Dylis (1966). *Fundamentals of Forest Biogeocoenology*, 634.94 (trans. by J. M. McLennan), Oliver and Boyd, Edinburgh and London.

Supan, A. (1896). *Grundzüge der physischen Erdkunde*, Leipzig.

Tanner, W. F. (1961). An alternative approach to morphogenetic climates, *Southeastern Geologist* 2(4), 251-257.

Tansley, A. G. (1920). The classification of vegetation and the concept of development, *J. Ecol.* **8**, 118-149.

Tansley, A. G. (1935). The use and misuse of vegetational terms and concepts, *Ecology* **16**, 284-307.

Tansley, A. G. (1949). *Britain's Green Mantle: Past, Present, and Future*. Allen and Unwin, London.

Tansley, A. G. (1953). *The British Islands and Their Vegetation*. Cambridge University Press (2 vols.).

Tardivo, Capt. (1913). *International Archives of Photogrammetry* (Vol. 6), International Society of Photogrammetry, Vienna.

Thornthwaite, C. W. (1948). An approach towards a rational classification of climate, *Geographical Review* **38**, 55-94.

Thornthwaite, C. W. and J. R Mather (1955). *The Water Balance*, Publications in Climatology, 8(1), Laboratory of Climatology, Centerton, New Jersey.

Thornthwaite, C. W. and J. R. Mather (1957). *Instructions and Tables for Compiling Potential Evapotranspiration and the Water Balance*. Publications in Climatology, 10(3), Laboratory of Climatology, Centerton, New Jersey.

Townshend, J. R. G. (Ed.) (1981). *Terrain Analysis and Remote Sensing*, Allen and Unwin, London.

Travaglia, C. and C. W. Mitchell (1982). *Applications of Satellite Remote Sensing for Land and Water Resources Appraisal, People's Democratic Republic of Yemen*, Tech. Report RSC Series 9, TCP/PDY/0104 (Mi), FAO, Rome.

Tricart, J. and A. Cailleux (1972). *Introduction to Climatic Geomorphology* (trans. from the French by Conrad J. Kiewiet de Jonge), Longmans, London.

Troll, K. (1943). Die Frostwechselhaufigkeit in den Luft und Bodenklimaten der Erde, *Meteorologische Zeitschrift* **60**, 161-171.

Troll, K. (1958). Climatic seasons and climatic classification, *Oriental Geographer* **2**, 141-165.

Troll, K. (1971). Landscape ecology (geoecology and biogeocenology), a terminological study. *Geoforum*, 43-46 (August 1971).

Tucker, C. J., C. Vanpraet, E. Boerwinkel, and A. Gaston (1983). Satellite remote sensing of total dry matter production in the Senegalese Sahel, *Remote Sensing of Environment*, 13, 461-474.

Tuomikoski, R. (1942). Untersüchungen über die Vegetation der Bruchmoore in Ostfinland, *Ann. Bot. Yanamo* **17**(1), 1-203.

UNESCO (1967). *Quaternary Maps of Europe, Africa, etc., Scale 1 : 2,500,000*, Paris.

UNESCO (1968). *International Tectonic Map of Africa, 1 : 5,00,000* in association with Association of African Geological Surveys (ASGA), Paris.

UNESCO (1970). *International Hydrogeological Map of Europe, 1 : 500,000*, Paris.

UNESCO (1973). *International Classification and Mapping of Vegetation*, UNESCO, Paris.

UNESCO (1974). *Metamorphic Map of Europe, 1 : 2,500,000*, Paris.

UNESCO (1975). *Legends for Geohydrochemical Maps*, Paris.

UNESCO (1976a). *Computer Handling of Geographical Data*, R. F. Tomlinson (Ed.), Natural Research 13, UNESCO Press, Paris.

UNESCO (1976b). *Engineering Geological Maps: A Guide to Their Preparation*, Paris.

UNESCO (1979). *Map of the World Distribution of Arid Regions, 1 : 25,000,000*, Paris.

U.S. Department of Agriculture, Regional Salinity Laboratory (1954). *Diagnosis and Improvement of Saline and Alkali Soils*, Agriculture Handbook No. 60, Riverside, California.

U.S. Department of Agriculture (1976). *Soil Taxonomy*, Agriculture Handbook No. 436, Washington, D.C.

U.S. Geological Survey (1970). *Apollo 6 Photomaps of the NW Corridor from the Pacific Ocean to Northern Louisiana, 1 : 500,000* (4 sheets), Washington, D.C.

van Beers, W. F. J. (1958). *The Auger Hole Method for Field Measurement of Hydraulic Conductivity*, International Institute for Land Reclamation and Improvement, Wageningen, The Netherlands.

Veatch, J. O. (1933). *Agricultural Classification and Land Typing of Michigan*. Michigan Agriculture Experiment Station, Special Bulletin No. 231.

Verstappen, H. Th. and R. A. Van Zuidam (1968). ITC system of geomorphological survey in *Use of Aerial Photographs in Geomorphology*, Chapter 2, Vol. VII, Delft, The Netherlands.

Warming, E. (1909). *Oecology of Plants: An Introduction to the Study of Plant Communities* (trans. by P. Groom and I. B. Balfour), Clarendon Press, Oxford.

Webster, R. (1977). *Quantitative and Numerical Methods in Soil Classification and Survey*, Clarendon Press, Oxford.

Webster, R. and I. F. T. Wong (1969). A numerical procedure for testing soil boundaries interpreted from air photographs, *Photogrammetria* **24**, 59-72.

Wilson, E. (1920). Use of aircraft in forestry and logging, *Canadian Forestry Magazine*, 439-444. (October 1920).

Wood, J. G. and R. J. Williams (1960). Vegetation. In *The Australian Environment*, Chapter 4, Melbourne University Press, pp. 66-84.

Wooldridge, S. W. (1932). The cycle of erosion and the representation of relief, *Scottish Geographical Magazine* **48**, 30-36.

Zieger, E. (1928). Ermittlung von Bestandmassen aus Flugbildern mit Hilfe des Hugershoff-Heydeschen Autokartographen, *Versuchsanstalt zu Tharandt* **3**, 97-127.

INDEX